The Good Solar Guide

Finn Peacock

RETHINK PRESS

First published in Australia 2018
by Rethink Press (www.rethinkpress.com)

© Copyright Finn Peacock

All rights reserved. No part of this publication may be reproduced, stored in or introduced into a retrieval system, or transmitted, in any form, or by any means (electronic, mechanical, photocopying, recording or otherwise) without the prior written permission of the publisher.

The right of Finn Peacock to be identified as the author of this work has been asserted by him in accordance with the Copyright, Designs and Patents Act 1988.

This book is sold subject to the condition that it shall not, by way of trade or otherwise, be lent, resold, hired out, or otherwise circulated without the publisher's prior consent in any form of binding or cover other than that in which it is published and without a similar condition including this condition being imposed on the subsequent purchaser.

Cover image © Mark Cavanagh

Contents

Introduction	1
My job is not to convince you to go solar	2
The genesis of mass market solar	3
How to use this book	5
STEP 1: Essential Knowledge	9
Fundamental 1: Solar energy versus solar power	10
Fundamental 2: Solar rebate versus solar feed-in tariff	19
Fundamental 3: The different types of solar system	22
Fundamental 4: Solar exports, self-consumption and your bill	33
Fundamental 5: How your roof affects your solar	41
Fundamental 6: The high cost of cheap solar	50
Fundamental 7: Don't get screwed over, know your rights	53
Summary	56
STEP 2: Measuring Your Energy Use	59
Energy audit step 1: Benchmark your gross daily usage	61
Energy audit step 2: Measure how much energy you use by day and by night	62
Summary	78

STEP 3: Heating Your Water — 79
- Your existing system — 79
- Your options for solar hot water — 84
- Summary — 93

STEP 4: Show Me The Money! — 95
- Sizing your system — 95
- The different approaches to buying solar — 98
- The costs of buying solar — 99
- Calculating your predicted payback — 106
- Getting more sophisticated — 106
- Do batteries make solar payback even better? — 110
- Summary — 115

STEP 5: Choosing Your Hardware — 117
- Panel-level optimisation (PLO) — 119
- Solar inverters — 121
- Batteries, inverters and surviving the zombie apocalypse — 128
- Solar panels — 132
- A monitoring system for your solar — 146
- Summary — 151

STEP 6: Getting Quotes — 155
- How to find a reputable solar installer (or three) — 161
- Site inspection — 168
- A comprehensive quote — 171
- Terms and conditions — 183
- Summary — 185

STEP 7: After The Install — 187
- Checking your install — 187
- Shifting loads — 191
- Checking your bills — 193
- Solar system maintenance — 197
- Summary — 199

CONCLUSION: Beyond Solar — 201
- Efficiency — 201
- Batteries — 202
- Cars — 202
- What not to do — 203
- A final note — 204

Acknowledgements — 206
The Author — 207

Introduction

I first entered the Australian solar industry in 2007. Back then, a grand total of 1,115 solar systems were in operation across the country and, by today's standards, most of them were tiny. Early adopters were paying around $10,000 for a four-panel solar array on their roof – that's $2,500 per installed panel – and systems would take 30 years to pay back their cost.

Fast-forward just ten years and there are over 1.7 million systems in use, with an average size of 15 panels. There are now more solar panels than people in Australia.

Today, in 2018, the approximate cost to install a big 5 kW system is $6,500. That's $325 per panel compared with $2,500 per panel ten years ago. At these prices, many people are paying the cost of their systems back in under four years and getting tiny bills along the way. To put it into perspective, in just ten years we've gone from 0.01% to over 17% of Aussie homes having solar.

If you're reading this, the chances are that you're one of the 83% of Australians who haven't yet installed solar. You no doubt know that Australia is one of the most sun-blessed countries on the planet. You can see that it makes a hell of a lot of sense to collect that sunshine and use it to power your home – especially given our increasingly hot summers, and with ever-rising grid electricity prices.

But you haven't put those panels on your roof yet.

My job is not to convince you to go solar

Let me put you at ease. Although I love solar (and as you read the rest of this book, I hope that becomes apparent), the aim of this book is not to convince you to go solar. There are plenty of books, blogs and company brochures that declare solar power the answer to all the world's ills. They imply that you must be a complete dunderhead (or worse, a planet-hater) if you haven't installed the magic panels on your roof yet.

This book is not like that.

I can't advise whether you should go solar or not because I have no idea what your situation is. I don't know how much energy your home uses, when you use it, what your roof looks like or how long you plan to be in your home. Solar is not for everyone, and it may not be for you.

My aim with this book is to give you all the information you need to decide *for yourself* whether solar makes sense and, if it does, show you how to buy a kick-ass system that will last for decades – at a great price.

If the simple analysis we do together shows financial returns that you are happy with then we'll get into the details of how to buy a great solar system, looking at:

- how many panels you really need
- whether you should get batteries too
- which panel and inverter brands to consider and which you should avoid
- how to find a great installer and a good deal, and
- how to make the most of your installed system to minimise your future bills

Introduction

The result should be that your home is blessed with tiny electricity bills for decades to come.

If you follow the steps in this book and you end up in any other situation, please go to Amazon and leave me a one-star review, or shoot me an email at finn@finnpeacock.com and we'll work out how to fix your situation together.

The genesis of mass market solar

In 2006, I was working for the Commonwealth Scientific and Industrial Research Organisation (CSIRO) in their Renewables and Energy Efficiency Division. Back then, as I've already pointed out, solar cost $10,000 for a tiny four-panel 1 kW system with a 30-year payback. Not the world's most compelling offer!

But in May 2007, six years after he created the Renewable Energy Target, Liberal Prime Minister John Howard (just before the election – go figure) announced an $8,000 rebate for anyone installing solar panels. At a stroke, Honest John had reduced the simple payback for solar from at least thirty years to a much more attractive five years.

Whenever free government cash is announced, business soon jumps in to help relieve the treasury of their crisp $100 bills as quickly as is humanly possible. Lo and behold, within weeks, from my ivory tower at CSIRO I started to see brand new solar companies beating a path to unsuspecting Aussies' doors offering to install a kilowatt of solar for prices as low as *minus* $500.

That's right: they gave you a $500 Myer voucher and a 1 kW solar system on your roof. All you had to do was surrender your $8,000 rebate to them.

And that was the birth of the mass adoption of solar in Australia.

I remember thinking two things at the time:

1. This market is likely to grow fast.
2. This market is going to have a problem with shonks selling crap to reap the generous rebate.

To cut a long story short,[1] I quickly left CSIRO and set up a solar website to give people advice when buying solar.

Fast-forward nine years and that website (SolarQuotes.com.au) has grown to become the most popular[2] solar information site in Australia. One in thirty homes in Australia has registered with us to get quotes for solar, and along the way I've answered tens of thousands of questions from solar buyers via email, Facebook and blog comments. More importantly, I've used this interaction with real Aussie families to develop a seven-step process to help Australians specify, buy and use a solar system that will lock in low bills for decades.

That seven-step process forms the basis of this book. My hope is that it can be the go-to guide for anyone installing solar, with the result that they pay a fair price for a great system that delivers tiny bills.

---

[1] Visit www.solarquotes.com.au/about for the full story.

[2] Ranked number one on Alexa.com for most visitors in Australia of any solar-specific website. Checked January 2018.

Introduction

How to use this book

I promise on the cover, in big letters, that I'll give you seven steps to tiny bills. This book is divided into seven sections – one for each step of the process.

STEP 1: Essential knowledge

This section goes over the solar basics that I think you need to know to understand what you are getting into with solar, and how it can best work to get your bills down. I don't take you any deeper than necessary – but if I do pique your interest and you want to get deep down and technical, I've provided links where you can join me online for an optional in-depth techno geek-fest.

STEP 2: Measuring your energy use

This section shows you how to measure your home's energy consumption so you can do your own energy audit. Buying solar without first performing an energy audit is like buying shoes without knowing your shoe size. In this section, you'll learn how to quickly measure your energy-use profile over 24 hours using nothing more than a pencil, paper and a clock. This is an important first step to working out whether solar would be a good investment for your home.

STEP 3: Heating your water

This section shines the spotlight on the (undeservedly) poor cousin of solar electricity: solar hot water. Most people give almost no thought to their water heating, despite the fact it is one of the largest energy users in most homes, second only to heating and cooling spaces. In this section, we'll work out how you currently heat your water, and decide whether to upgrade to heating it with the sun instead of grid electricity or gas.

STEP 4: Show me the money!

This section shows you how to accurately estimate the financial return of a solar system on your roof and goes through the economics of batteries. For most of us, money is a scarce resource. Only the most committed greenie will add solar to their home if the financial return is not there. Here, you get to understand in full the financial implications of solar (and batteries) without the pressure of a salesman in your home. And because you're using data you collected from your meter in Step 2, you can be confident you will get close to this return when your first post-solar electricity bill arrives.

If, after Step 4, you decide the payback is good enough, then in Steps 5 and 6, I will walk you through specifying and buying a solar system that will make good on that return on investment by providing the energy you need for decades to come.

STEP 5: Choosing your hardware

A big part of buying a solar system is choosing what hardware to use. Most people think that the most important decision is about the solar panel brand. But there is an even more critical choice to make, and this concerns the box of power electronics called the 'solar inverter'. This section walks you through choosing an inverter that will have the reliability and features that you need, then moves on to the wonderful world of solar panel brands. There are heaps of brands available in Australia, many of them cheap and nasty. I'll show you the two dozen or so that you should

Introduction

consider if you want a safe, reliable and well-supported system on your roof. To round things off, I'll dive into a long-neglected piece of the solar puzzle – third-party monitoring – and explain why I consider it an essential addition to any solar system.

STEP 6: Getting quotes

Once you understand your hardware choices, you can get quotes safe in the knowledge that you can have an intelligent conversation with the potential installers. This section shows you how to find three reputable installers, engage them for quotes and then pick the best quote for you. Having been through Step 5, you'll understand that getting the cheapest system on the market is a really bad idea, but you'll still want a system that's good value. This section shows you how to navigate the Australian solar market to get high-quality, good-value solar fitted by conscientious installers. It also shows you what information your quote should include. I'll share my insider secrets with you from ten years in the industry.

STEP 7: After the install

Any law-abiding installer will fit your system to the current standards and guidelines. A good installer will go over and above the guidelines and give you bulletproof confidence in your new solar power station. Here, we go through a post-install checklist of small but important things that are not necessarily in the standard but should still be done to ensure your system has a long and hassle-free life.

Before parting, I'm also going to show you how to live with your new solar system so you keep your bills as small as possible. But fear not, you don't need to live like a hermit. By the time you get to this section, you'll have a high-quality power station on your roof that will provide the security of low bills for decades to come. And we'll have fun getting there, because living the good life with tiny bills feels pretty damn fine from my experience. My last summer power bill for a family of five and a small office was a $128 credit for the quarter. Let's see how low we can get your bills!

Sounds like a lot of work? It's not, really – plus, it's worth it when you consider this: over the next 30 years, a typical Aussie household like yours is likely to save over $40,000 in today's money by filling the roof with panels. This book will take you about an hour to read. I reckon even Donald Trump could be persuaded to sit down with a book for an hour in return for forty grand...

Motivated? Good. Let's get started.

STEP 1: Essential Knowledge

Imagine buying a new car without knowing the difference between distance and speed. Without knowing that you buy fuel in litres. Without knowing that diesel and petrol and LPG are different fuels. Without understanding the difference between a ute, an SUV, a wagon and a hatchback. Without realising that paying more for a Hyundai than a Ferrari is a really bad deal.

You'd be completely at the mercy of the car salesperson. The sad truth is that many car salespeople are more interested in selling you the car that makes them the biggest commission, or in shifting stock, than in finding the one car that will best serve you for the next few years. And even if you luck out with a salesperson who will scour the earth to find you the right car, they certainly aren't going to recommend a brand that they don't sell.

Can you imagine a Holden dealer saying, 'Actually, the Ford Territory would be a great car for your family – pop over to the Ford dealer and they'll look after you'?

You should also know what kind of fuel economy it is reasonable to expect. If you just walk from dealer to dealer demanding the most efficient car, you're unlikely to get the car that suits you. Yet people will often demand the most efficient solar panel – not realising they could be paying a 100% premium for a panel that's only 5% more efficient.

You also need to know some background to decide what extra features you need and have a good idea of what the whole package should cost.

When you're buying a new car, you need a base level of knowledge. This is knowledge that you've no doubt picked up already because cars are such a big part of Western culture. However, solar is such a new industry that you need to learn the basics before buying.

By the time you've finished this section, you'll have a really good feel for the solar basics. It's a good investment of your time because solar is going to play a big part in the world's energy future. People who don't understand the basics will be the ones left complaining while they pay through the nose for grid electricity or struggle with a pile-of-crap system on their roof that's nothing but a huge pain in the backside. Meanwhile, the knowledgeable homeowners, like you, will enjoy clean, safe, cheap, reliable energy from the sun.

Fundamental 1: Solar energy versus solar power

What is the difference between 'solar energy' and 'solar power'? Wait, you thought they were one and the same thing?

Nope. Believe it or not, there's an important difference between solar power and solar energy. To be more specific, there's a large and important difference between what constitutes power and what constitutes energy.

There's a reason I'm starting the book with this weird-sounding question. To specify a solar power or battery system for your home, you absolutely need to understand the difference. If you don't, you can get into all sorts of trouble.

STEP 1: Essential Knowledge

Let's go back to the car analogy. Imagine you were buying a car and you didn't know the difference between speed and distance. Sounds ridiculous, right? You might ask for a car that can go 800 kph between filling up (when you really need a range of 800 km). Or a car that can go from 0 to 60 km in under 5 seconds (when you really mean 0 to 60 kph). Or imagine telling the salesperson that you'll never drive over 110 km because you stick to the speed limit.

It all sounds ridiculous and it's a recipe for total confusion. You'd be highly unlikely to get the right car for you if you didn't know the difference between speed and distance. Yet this situation happens every day in Australia when people are buying solar. And as home batteries go mainstream, it will only get worse. There was a *Catalyst* special on ABC recently and even they got their power and energy mixed up. God help us.

Read on. In about four minutes, you'll fully understand the difference between power and energy and thus will know more than the average *Catalyst* presenter, as well as, I'm sad to say, some solar salespeople.

Here's the first thing you need to know:

- Power is measured in kilowatts (**kW**)
- Energy is measured in kilowatt-hours (**kWh**)

Let's break those terms down:

- **k** stands for kilo, which means one thousand
- **W** stands for watt, which is a measure of power
- **h** stands for hour, a measure of time

What is power?

Power is the rate at which energy flows. Here are some examples when talking about electricity.

- The rate at which a solar panel can push out electricity in full sun, for example, 'a 250 W solar panel'
- The maximum rate at which a battery can accept electrical energy when charging, for example, 'the battery can charge at 3 kW'
- The maximum rate that a battery can kick out its stored energy when discharging, for example, 'the battery can discharge at 3 kW'
- How quickly an appliance gobbles energy, for example, 'that heater uses 2 kW'

If you think of electrical current as water in a hosepipe, its power is the flow rate at which the water travels.

Figure 1.1 Electrical power is analogous to the flow rate of water through a hosepipe.

The size of a solar system is defined by the 'peak power' in kW, of its solar array (where 'solar array' is the collective term for all the solar panels). For example, a 3 kW solar system, might consist of

STEP 1: Essential Knowledge

ten 300W solar panels on the roof. This solar array can push electricity out at a maximum rate of 3 kW (3,000 watts).

For most of the day the solar panels will not produce at their peak power. Only in full midday summer sun, in perfect conditions and with perfectly clean panels, will the electricity flow out of those panels at the system's nameplate peak power. For example, that 3 kW solar array should give out 3 kW of power under perfect conditions.

Figure 1.2 The power from a 3 kW solar system changes throughout a continuously sunny day.

In practice, you often get about 20% less than the peak power rating because of unavoidable losses in the system, such as:

- dirt on the panels
- the resistance of the wires to your roof
- solar inverter losses
- temperature losses from the solar panels (more on this later)

So the curve for a 3 kW system in the real world typically peaks at closer to 2.4 kW.

> **Online resource:** Find more info about solar panel losses here: solarquotes.com.au/losses

To recap, a solar system's size is defined by the peak *power* output of the solar array.

What is energy?

The abbreviation kWh stands for kilowatt-hour. A kWh is a measure of energy (not power). Energy is how much electricity has been generated, stored, or consumed over time.

With our water analogy, where power is analogous to flow rate, if the flow is directed at a bucket, the amount of water collected in the bucket is analogous to electrical energy.

Figure 1.3 The water/electricity analogy.

If your solar panels (for example) continuously give out 5 kW of power for a whole hour, you will have produced 5 kWh of energy.

STEP 1: Essential Knowledge

That energy could get used by your appliances, it could be exported to the grid or, if you want to get fancy, it could be stored in a battery – just like storing water in a bucket. Or it could be divided among the three.

The amount of electricity you use (or generate or store) is defined in kWh. For example, 'My solar system produced 4 kWh of electricity today!' or 'My heater used 2 kWh of electricity today' or 'This battery can store up to 10 kWh of energy'.

Key point: A good way to cement the difference between power and energy is the battery and bucket analogy. With a bucket and a hose, the higher the power of the hose, the faster you can fill your bucket. The size of the bucket determines how much water the bucket can accept.

With a battery, the higher the power rating of the battery (in kW), the faster you can fill or empty the battery. The higher the energy capacity (in kWh) of the battery, the more electricity it can store.

When you get billed for electricity, you get billed for how many kWh of energy you have used.

At the highest level, kW measures power, and kWh measures energy.

If we look at the solar system power curve graph again, the amount of energy generated in kWh is shown as the area under the curve:

Figure 1.4 The energy generated by a solar system in one day is the area under the power curve.

Why is the difference between energy and power important?

It is common for people to mistakenly interchange the terms energy and power as if there were no difference. Most people do it all the time without noticing, even many electricians. It drives electrical geeks like me up the wall. Especially when I read it in national newspapers and books!

For example, if someone is talking about their electricity use and says, 'I used 8 kW yesterday,' strictly speaking, they mean, 'At one point in time yesterday my power demand got to the point when I was pulling 8 kW of power from the grid.' It gives us no indication of how much *energy* they actually used – only the peak *power* they hit.

It is the equivalent of saying, 'I went for a drive and did 140 kph yesterday.' It gives us no idea how much distance they travelled, only that, at least once, they hit a max speed of 140 kph (they must have bought the Ferrari, not the Hyundai).[3]

---

[3] OK, OK, Hyundais can totally do 140 kph. All that's required is a cliff.

STEP 1: Essential Knowledge

The person who thinks they used 8 kW almost certainly means that they used 8 units of electrical energy yesterday, in which case they should have said, 'I used 8 kWh yesterday.'

Yeah, yeah, I know what you're thinking: who cares?

Well, it's like our driver saying 'I went 140 kph' when they actually meant 'I went 140 km'. It totally changes the meaning, and causes ambiguity that could be costly.

This is important if you're buying a solar system. If someone says they need a solar power system to produce 8 kW, they might end up being quoted for an 8 kW solar system. That will cost about $10,000 at today's prices and produce about 32 kWh of energy per day averaged over a year.

Figure 1.5 An 8 kW solar system is much bigger than an 8 kWh solar system.

17

Please don't confuse kW and kWh. If you do, you may end up with a solar system that's completely the wrong size.

If what they actually meant was that they needed to cover an energy use of 8 kWh per day then they really need a 2 kW solar system, which costs about $3,000 at the time of writing and produces, on average, 8 kWh of energy per day.

> ### Tip
> A simple way to estimate how much energy you can expect, on average, per day from a solar system in Australia is to multiply the system size by 4. For example, a 5 kW system will average around 20 kWh of energy production per day.
>
> ### Top tip for filtering out the worst solar salespeople
> Ask them to explain the difference between a kW and kWh. If they get this wrong, how on earth will they understand your requirements? A lot of cold-calling door-knockers and solar telesales people will fail this test.
>
> In fact, never entertain a cold call by phone or at the door. There are just too many crooks out there.

Online resource: If you come across any other jargon on your solar journey, I've got a glossary of solar terms here:solarquotes.com.au/glossary

Now you know the difference between solar power and solar energy (or, more generally, the difference between power and energy), you are well equipped for a future where you take more responsibility for your own energy. It's like knowing what a

STEP 1: Essential Knowledge

gigabyte is – unthinkable ten years ago, indispensable now if you want to understand whether your next smartphone will store all your apps, photos and videos.

Fundamental 2: Solar rebate versus solar feed-in tariff

When it comes to the solar rebate and your solar feed-in tariff, in my experience, most people outside the solar industry think these two things are one and the same. In fact, apart from mixing up kW and kWh, it's the biggest misunderstanding I come across when talking to people who are considering going solar.

The classic line I hear is: 'The rebate has gone way down to virtually nothing! Solar's not worth it – the opportunity has passed!'

This belief couldn't be more wrong. To help you understand why, let me explain exactly what these two terms refer to.

The solar rebate[4]

If you buy a solar system today, it is subsidised by a government-administered scheme (small-scale technology certificates, known as STCs). At the time of writing (early 2018), this scheme will save you about $600 per kW of solar installed. That's a saving of around $3,000 for a typical 5 kW system.

> **Online resource:** See the current maximum solar rebate value where you live: solarquotes.com.au/rebate

---

[4] The government doesn't want us to call it the solar rebate. They want us to call it the solar financial incentive. In the real world almost everyone calls it the solar rebate, so I will too.

The rebate is applied at the point of sale (meaning any advertised prices you see almost certainly have the rebate already applied).

Under the current legislation, this rebate dropped by one-fifteenth in January 2017, one-fourteenth in January 2018, and it will drop every January until it reaches zero in 2031.

The solar rebate subsidises the upfront cost of installing a solar power system and is not means-tested in any way. The only criteria for claiming it are:

1. Your system is less than 100 kW in size.
2. You get it installed and designed by a Clean Energy Council-accredited professional.
3. You use panels and inverters that are approved for use in Australia by the Clean Energy Council.
4. You are not replacing existing panels in an existing system.

Interestingly, these rules have all but removed the concept of DIY solar in Australia. If you are not accredited and do the installation yourself (though, legally, you'd need a sparky for the 230 V parts), you can't claim the rebate. The rebate should easily cover the cost of installation, though, making it cheaper to get the professionals in.

The best things about the solar rebate are that everyone is eligible, you can claim it as many times as you want (up to 100 kW per system) and you don't have to do anything to claim it except engage a good installer who will process all the paperwork for you.

The feed-in tariff

The feed-in tariff is something completely different from the rebate, so don't get them mixed up!

STEP 1: Essential Knowledge

When your rooftop solar generates more electricity than your home is using at any point in time, it isn't wasted – instead, it's sent into the grid for other people to use. A feed-in tariff is the payment you receive for selling electricity to the grid.

Although feed-in tariffs are still available (and should always be), they are worth much less than they used to be. Hence the mistaken belief that the rebate has dropped. The rebate is very much alive and kicking, it's the feed-in tariff that's dropped.

Feed-in tariffs were initially high to encourage the uptake of solar, back when solar cost three to four times the price it is now. It worked. The solar industry is well established and prices for Australian solar systems are the lowest in the world, thanks to high-volume and efficient installers.

Unfortunately, outside the Northern Territory and some parts of regional Western Australia, feed-in tariffs for new rooftop solar are low compared to the retail price of electricity.

The quickest and easiest way to find the solar feed-in tariffs currently offered by electricity retailers in your area is to go to my online electricity retailer comparison page and enter your postcode.

> **Online resource:** Compare feed-in tariffs in your area: solarquotes.com.au/energy

This tool compares most of the electricity retailers in your area and you may well find you can get a much more generous feed-in tariff by shopping around. Look at typical feed-in tariffs and consumption tariffs in your postcode and write them down either here or in the downloadable worksheet that covers all the exercises in the book at solarquotes.com.au/worksheet. We'll use them when working out the payback of solar.

Typical feed-in tariff in my postcode	c per kWh
Typical usage tariff in my postcode	c per kWh

The important thing to understand is that this feed-in tariff is not the same as the rebate. Feed-in tariffs have come down from their highs of five years ago, but the rebate is still generous.

Fundamental 3: The different types of solar system

In May 2015, the public's perception of solar power changed overnight. What happened? Billionaire Elon Musk jumped on stage in California and announced a product called the Tesla Powerwall, a sleek-looking home battery.

For people living off-grid, the concept of powering your home with a battery had been around for decades, but Musk's gift for publicity took the concept from the off-grid fringe to the grid-connected mainstream. Overnight, solar installers were bombarded with people wanting batteries with their solar.

I'll talk about the pros and cons of adding batteries to your system later. Right now, I want to quickly go over the different types of solar systems and where batteries come into the equation.

At a high level, there are three types of solar system:

1. On-grid solar.
2. Off-grid solar.
3. Hybrid solar.

Let's go through each option briefly.

STEP 1: Essential Knowledge

On-grid solar

On-grid solar is also known as:

- grid-connect solar
- grid-tie solar, and
- grid-feed solar

This is still the most common solar system by a country mile. Ninety-five per cent of solar systems in Australia are of this type.

This is a solar system that is connected to the grid. It has no batteries connected to it.

Figure 1.6 shows the concept:

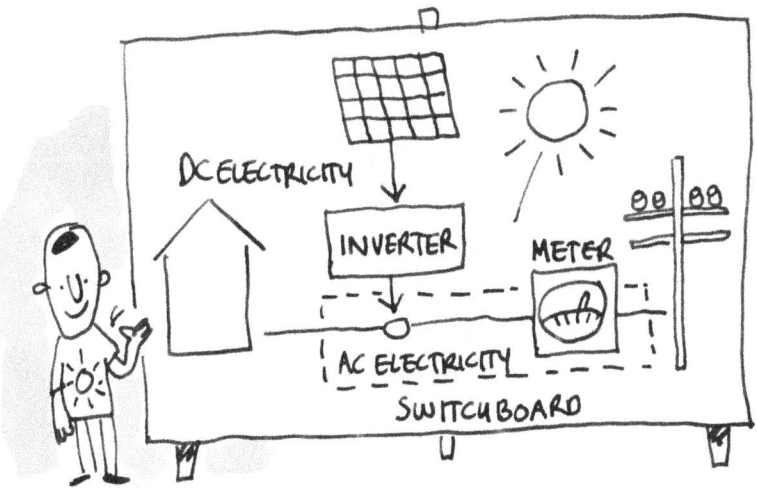

Figure 1.6 An on-grid solar system, no batteries.

The solar panels generate direct-current (DC) electricity when light hits them.

Remember, it is the light from the sun that generates the electricity, not the heat. Heat actually reduces the panels' efficiency, as we'll learn a little later.

How do solar panels generate electricity from light? The light knocks electrons about in the silicon wafers that make up the panels. Those electrons are caught by tiny wires called 'busbars', which are laid on the silicon. This process is called the 'photovoltaic effect'.

> **Online resource:** How the photovoltaic effect works: solarquotes.com.au/pv

The DC power is 'direct current', which means it is a steady voltage and current. This voltage can be high: up to 600 V in a residential installation.

But the appliances in your home don't use DC; they use 'alternating current' (AC). AC means that the current wiggles up and down 50 times a second. The reason they use AC power is that, at the advent of large-scale electricity generation, AC was much easier to generate. Why? Because all generators were made to spin. For example, a steam turbine uses steam to spin a generator. A spinning generator without any modern power electronics to smooth it naturally generates AC as it spins round.

Also, AC is much easier to put through transformers to jack up the voltage to hundreds of thousands of volts. Then the current can be efficiently transmitted long distances from power stations (or wind farms) to your local substation.

The whole developed world is set up for AC power, so we need to convert the DC solar power to AC power, which in Australia is 230 V AC, cycling at 50 times a second. That DC to AC conversion

STEP 1: Essential Knowledge

is done by the solar inverter. The solar inverter is a box of power electronics that sits on your wall. It converts solar DC to usable AC, which is fed directly into your home's switchboard.

From the switchboard the solar power will first flow into any appliances in your home that are using power. There will always be some electricity consumption in a modern home, so whenever there is solar generation, at least some of it will flow into the house.

The on-grid solar system has two basic modes of operation, which depend on how much solar is being generated and how much electricity your home is using.

Mode 1: Surplus solar

If there is more solar going into your switchboard than your appliances can use at any point in time, the excess solar will simply be exported to the grid.

Figure 1.7 Surplus solar flows into the grid.

This excess solar (Arrow C in Figure 1.7) flows through your meter recording how much power is flowing out. The meter counts how many kWhs go out into the grid. It keeps tally on one of the digital counters that you can scroll through on your meter's liquid crystal display (LCD). Your electricity retailer (the company that bills you every quarter) will record this count in every billing cycle (usually every three months). If you have a 'smart meter', they get the number over the air. If your meter is not smart, someone comes and reads it manually. The retailer will pay you for that exported electricity.

Mode 2: Not enough solar

If, at any point in time, you are not generating enough solar for your appliances to use, your switchboard imports grid electricity to make up the shortfall, as shown in Figure 1.8.

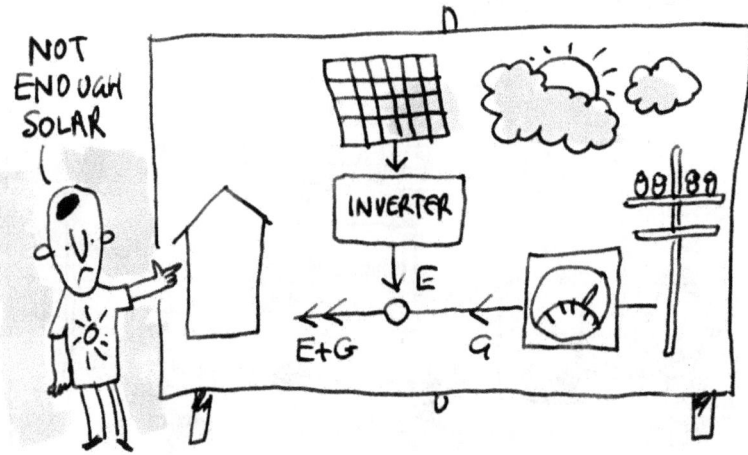

Figure 1.8 Not enough solar so electricity is imported from the grid to top it up.

STEP 1: Essential Knowledge

The energy from the grid in Figure 1.8 flows though the meter too. The meter records how much grid energy you import so you can be charged for it.

Again, the meter cannot measure your home's total electricity consumption (which is the sum of E and G in the diagram). It can only measure your grid imports (G). You'll need to buy your own monitoring if you want to see what's going on 'behind the meter', and I'll show you how to do that in Step 5: A monitoring system for your solar.

The concept of 'behind the meter' You expect your electricity meter to know how much energy you use. Sounds obvious, right?

But generating your own electricity needs a whole new mindset. You may be surprised to learn that the meter installed by the electricity retailer doesn't know – and can't know – the details of what is happening with your home's electricity.

When you have surplus solar, your meter can't see how much electricity your home is using or how much is being generated by the solar system. It can only measure the exported solar.

For example, if you're exporting 2 kW of surplus solar your meter doesn't know if you're generating 3 kW and using 1 kW, or if you're generating 4.3 kW and using 2.3 kW. All it knows is that the difference is 2 kW.

Figure 1.9 shows the physical layout of a grid-connect system.

The Good Solar Guide

Figure 1.9 An on-grid solar system.

Off-grid solar

An off-grid solar system is... drum roll, please... not connected to the grid. But you already knew that.

What you may not appreciate is that an off-grid system that will give a typical Australian home almost the same convenience as being on-grid will cost at least $40,000. Why so much? Because it needs to be designed to work even in the depths of winter when there is very little sun. This means it needs a much bigger solar array to generate as much power as possible in low light and to charge your batteries as well as powering your home. It also needs a large battery bank to store as much energy as possible for the gloomy days and to cope with any appliances that draw high power, even for short periods.

An off-grid system is similar to an on-grid system, except that you need to add the following.

STEP 1: Essential Knowledge

- **A battery inverter:** Like solar panels, batteries produce DC voltage. The battery inverter converts this to AC for your appliances to use. The battery inverter also houses the power electronics that constantly balance supply and demand in the stand-alone system.
- **A battery bank:** Any solar power generated above your appliances' needs, instead of going to the grid, charges your batteries. Once the batteries are fully charged, the solar power is throttled back so as not to damage the batteries. This means that, most of the time, your solar panels are running way under their peak capacity – wasting solar energy, if you like. When your solar panels are not producing enough power, the appliances draw power from the batteries.
- **A generator for backup:** There will always be times when your batteries are drained and the sun refuses to come out. For these times, you need a backup generator.

Here are four reasons you might choose to go off-grid.

1. **There is literally no electricity grid where you live.** Perhaps you live in the middle of the Simpson Desert like my Uncle Dave (seriously, he does!) or somewhere else equally remote. Hey, there's no shortage of places like that in this big ol' country of ours!
2. **There is a grid, but the nearest connection point is a long way from your home** and your local electricity network wants to charge you an arm and a leg to connect you. In this case, it may be cheaper to go off-grid. Be aware, though, that a decent-sized off-grid system for a very efficient home is going to start at about $30,000.

3. **You are connected to the grid but you suffer frequent blackouts that cause you grief.** You want a system that can run when the grid is down and you understand that a standard grid-connect solar system doesn't work when there's a power cut, unless it has a big and expensive battery backup system or a backup generator (or both).
4. **You're crazy.** Your house is already connected to the grid. Blackouts aren't a problem (if they were, you'd already have a generator), but you just like the sound of going off-grid. You think it makes you more independent and protects you from the forthcoming apocalypse. In fact, you've already got a big shed to put the batteries in. It's the same one where you stored all the canned food in readiness for the Y2K bug/end of the world last time round. You don't mind spending $30,000 on an off-grid system that would only cost $6,000 if it were grid-connected. Or maintaining it much more frequently than a grid-connect system. Or buying new batteries every few years. Because you're much closer to nature now you've cut that grid connection…

I guess what I'm trying to say is that unless 1, 2 or 3 above applies, it's economically and environmentally insane to insist on an off-grid solution. It's economically insane because the cost is three to five times higher. It's environmentally insane because of all those batteries you'll need to buy (and replace from time to time), which may contain nasty chemicals. Batteries also reduce the efficiency of your solar system. Expect them to lose 10% to 30% of all the energy stored. There's also all the solar energy that's wasted when your batteries are full. With no grid to absorb excess solar energy your solar panels will spend most of their time being throttled so they don't overload your house and batteries.

STEP 1: Essential Knowledge

Luckily, there's a compromise: hybrid systems.

Hybrid solar systems

Hybrid solar systems are the best of both worlds: you get the guaranteed (well, 99.9% of the time) electricity supply of the grid, with the ability to store your excess solar energy for use when the sun isn't shining. This typically reduces your dependence on grid imports by 70% to 95%.

Hybrid systems are also at least half the price of an off-grid system and don't require diesel backup. They're still more expensive than a purely on-grid system, though – typically double the price of grid-connect solar.

The only difference between a hybrid solar system and a regular grid-connect system is the addition of batteries and a battery inverter – see Figure 1.10.

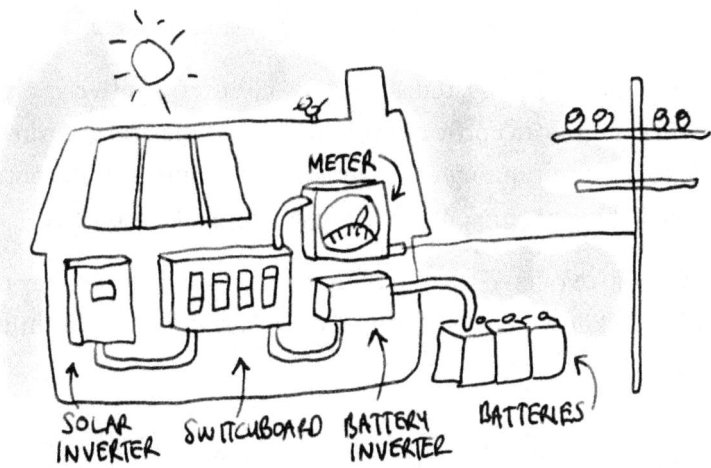

Figure 1.10 Hybrid solar system is an on-grid solar system with batteries.

A hybrid system is generally configured to charge the batteries with excess solar energy – as this is the cheapest form of energy you can get. Once the batteries are full, the excess solar gets sold to the grid, earning you an income.

When there is not enough solar energy to power your home, the battery inverter will do its best to provide power from the battery. If it can't provide enough power (remember, power is measured in kW) because it hits its power rating, or it cannot provide enough energy (kWh) because it is running flat, then the grid will provide the shortfall.

Generally, the batteries in hybrid solar systems are sized to get you through the night, but they can still pump out power during the day when required. Hybrid systems are configured so that your house uses solar first, then battery power, then – as a last resort – grid power. If you suffer a grid outage, some hybrid systems can provide limited backup from your batteries to keep the lights on and some appliances running.

I chose my words very carefully in the previous sentence.

I say 'some' hybrid systems because, counterintuitively, many hybrid systems can't provide backup power if the grid goes down. This is because running in 'off-grid' mode requires extra control and switching equipment, which costs at least $1,000 more.

Even if you're offered a battery that promises 'backup', it may not be the backup you're expecting. I'll explain this in detail in the battery section of Step 5.

In summary, a hybrid solar system is a grid-connect solar system with added batteries. It's about a third of the price of an off-grid system, can sell excess solar instead of wasting it, and doesn't need a generator.

STEP 1: Essential Knowledge

If you want batteries and you already have a grid connection (or you're building and you can get connected to the grid for a reasonable price), hybrid is a much better option than off-grid.

So now the question becomes: 'Should I buy a standard grid-connect system or a hybrid?'

Short answer: If your main motivation is optimal payback or helping the environment then don't buy batteries (yet).

Online resource: Battery myths: solarquotes.com.au/batterymyths

Fundamental 4: Solar exports, self-consumption and your bill

Have you heard about people getting solar and then basking in the joy of watching their meter run backwards?

The problem with this common anecdote is that, although it does happen with the old-style meters, it is (unfortunately) illegal.

When you get grid-connected solar, you need to get a new electricity meter installed. Your new meter is electronic, not mechanical, and it will not run backwards. Sorry.

In most jurisdictions, it is illegal to switch your solar system on before this new meter is installed. Even though it's tempting!

If a solar system is wired into an old 'spinning dial' meter then the meter will run backwards whenever the house is exporting solar electricity to the grid. The old meters are mechanical devices and when the electricity flows backwards – out to the grid instead of in from the grid – the mechanics go backwards too.

The Good Solar Guide

If it were not illegal, I would recommend everyone on solar get an old, mechanical meter. If you could simply run your meter backwards then you would reduce your usage by one unit of electrical energy (1 kWh) for every unit you exported back to the grid. Or, to put it another way, if you paid 30c for importing a unit of electricity from the grid, every time you exported a unit, you'd get that 30c back. Getting a zero-dollar electricity usage charge would be as simple as making sure your gross solar system generation over a year was the same as your home's gross electricity usage – the two would cancel out.

But that's not how it works.

In Australia, the price you get for excess solar energy exported to the grid is the feed-in tariff we've recently discussed. The minimum feed-in tariff in each state at the time of writing is shown in Table 1.

TABLE 1: Minimum feed-in tariffs

State/Territory	Mandated minimum feed-in tariff per kWh
NSW	No mandated minimum
ACT	No mandated minimum
VIC	9.9c
QLD (regional)	10.102c
QLD (south-eastern)	No mandated minimum
WA	Varies from 0c to 48c depending on location
NT	25.67c
TAS	8.929c
SA	6.8c

STEP 1: Essential Knowledge

The good news is that, if you have electricity retailer competition (i.e. everywhere except the Northern Territory, Western Australia, Tasmania, and rural Queensland), you can get a better feed-in tariff by shopping around. For example, in South Australia, at the time of writing, you can get up to 18c – more than double the mandatory minimum. Table 2 shows the best feed-in tariffs in each part of Australia at time of writing.

TABLE 2: Best feed-in tariffs by state

State/Territory	Best retailer feed-in tariff per kWh
NSW	17c
ACT	15c
VIC	12c
QLD (regional)	10.102c
QLD (south-eastern)	16c
WA	Varies from 0c to 48c depending on location
NT	25.67c
TAS	8.929c
SA	18c

The thing is, even earning 18c per kWh is a lot less than the 25c to 30c per kWh most people pay for electricity. And a lot of people get mighty upset at the discrepancy.

I think it's perfectly fair that we earn less for exports than we pay for imports. That's because the prices paid to us reflect the wholesale price of electricity.

Wholesale electricity prices

In all industries, you get wholesale and retail pricing. When a consumer (who is used to paying retail prices) gets exposed to the true wholesale price of the goods they are buying, they can get mighty upset. For example, my mate has a bicycle business and gets all his bikes wholesale. He can afford twice as good a bike as I can, because retail mark-ups in the bike industry are about 100%.

The price you pay for electricity is the retail price, which is much higher than the wholesale price – typically, three times more. This is because before you're billed, the retailer has to add on:

- transmission costs (for access to the long-range transmission network of poles and wires)
- distribution costs (for access to the local network of poles and wires)
- wholesale electricity costs
- a retailer margin
- Goods and Sales Tax (GST), and
- other surcharges mandated by the government

The result is that, on average, the wholesale cost of electricity is about 8c per kWh. If you're getting close to that, you're in the ballpark of fairness.

Yes, some of your electricity is going straight into your neighbour's house and bypassing most of the network, but for the system to work as a whole, society (including grid-connected solar owners) has to cover the cost of the entire system.

Victoria held an enquiry in 2016 to find a fair price for solar. In my opinion, it was a fair enquiry, and the result was that they upped the mandated minimum feed-in tariff to 6.5c to 8c

STEP 1: Essential Knowledge

depending on the time of day the electricity gets exported. In 2017 it was increased again to 11.3c. Whoopie do! Savvy solar owners have been getting that – or more – for years, simply by shopping around. Once again, the pollies are simply playing catch-up with the real world of market forces.

The point of all this – whether you get your knickers in a twist about 'unfair' feed-in tariffs or not – is that they are what they are. The odd enquiry may put them up or down by two or three cents, but if you want low bills in the meantime, you need solar, and if you want solar, you need to know how to work with low feed-in tariffs and still get tiny bills. It is possible, and easy.

Net metering

When you install solar, the amount of solar electricity you export will need to be measured so you can get paid for it. You'll either get a new meter or get your existing meter reconfigured. The new meter is called a 'net meter' (note: this is different from a smart meter). It has two counters:

Counter 1: Imports. This shows how much energy you import from the grid. You import from the grid when your home is using more than you generate with your panels at any moment in time. You pay the retail electricity rate for every kWh of electricity you import. This price can vary from 22c to 55c per kWh depending on your tariff.

Counter 2: Exports. This shows how much solar energy you export to the grid. You export when your panels are generating more solar than your home can use at any moment in time. You get paid a feed-in tariff for every kWh of electricity you export. As mentioned, this is generally from 6c to 16c per kWh depending on the retailer.

Net metering and zero-dollar bills

I've already covered how, if your meter ran backwards whenever your system exported energy, then it would be really easy to get a zero-dollar bill with solar.

Here's all you'd have to do. Before going solar, you'd take your most recent bills and look how many kWhs you'd taken from the grid in the last 12 months. Then you'd buy a solar system that would generate that many kWhs in 12 months.

Predicted solar generation in Australia is easy to calculate. As a rule of thumb, you generate about 4 kWh per day for every kW installed on the roof. The magic multiplier is 4.

Let's say your home used an average of 20 kWh per day. You could buy a 5 kW system to generate the same amount:

$$5 \text{ kW} \times 4 = 20 \text{ kWh per day}$$

Voila! Your net consumption from the grid would be zero.

You'd still have a service charge of about a dollar a day, but you could simply add another kW to the system size to offset that by generating another 4 kWh per day, which, at 25c per kWh, would reduce it to nothing.

Bingo – a zero bill.

Unfortunately, we can't do this (though some shonky solar companies are still calculating payback this way).

We can't do this because, in almost every home, a proportion of your solar energy will always be exported. And that exported energy is worth less than the price you pay for imports.

STEP 1: Essential Knowledge

This means your solar system is at its most valuable when you are consuming your solar energy yourself. When the solar you are generating is being used up in your home, you are avoiding buying energy from the grid, saving around 30c per kWh. Compare this to exporting at 6c to 16c per kWh.

Key point: The more solar you self-consume, the lower your post-solar bills will be and the faster your system will pay you back.

This is why it's important to measure not only **how much** electricity you use but also **when** you use it. Ideally you want to know how much energy you consume in at least 30-minute intervals. I call this your 'energy profile'.

Unfortunately, many solar installers do not measure this.

Now you understand how important your energy-use profile is to system payback, you won't be tempted to buy a system without first measuring your profile. You are on the first step to being an informed solar buyer with a realistic expectation of how much solar can help you. I'll show you how to measure your energy profile in Step 2.

First I need to expose how the electricity retailers hide the true value of solar when they send you your bills.

Your post-solar bills will hide most of your solar savings

When you get your first post-solar bill, the amount you have to pay will be the difference between the cost of your solar imports and your earnings for solar exports, plus your daily service charge.

39

The daily service charge is also known as a 'supply charge'. It's the amount you are billed, each day, for simply being connected to the grid. Typically, it's about a dollar a day.

So you pay:

> Usage − Exports + Daily Service Charge = Your Bill

Do you remember how I explained that your electricity retailer doesn't know what goes on 'behind the meter'? None of the solar that has been self-consumed, saving 25c to 50c per kWh, is shown on your bill. For most solar owners, that's the bulk of their savings!

This leads to many solar owners thinking that the only financial benefits from their systems are the exports on their bills. Typically that's $150 per quarter, or only $600 per year.

If the electricity retailers cared about their customers knowing what's really going on, they could plug a $20 device called a 'current transformer' into your meter to measure your solar output. Then they could present a bill that showed all your savings.

Let me use one of my bills as an example to show you what a difference that would make.

What my bill told me my savings were in a quarter:

> Solar exports: 2,457 kWh x $0.08 = $196

What my true savings were:

> Solar exports: 2,457 kWh x $0.08c = $196
>
> Solar self-consumption: 960 kWh x $0.31= $297
>
> Total solar savings = $493 for the quarter

STEP 1: Essential Knowledge

As you can see, accounting for self-consumption is important to see the true savings of solar.

In **STEP 4: Show Me The Money!** I'll show you exactly how I accounted for the self-consumption.

Fundamental 5: How your roof affects your solar

Assuming you buy a good-quality solar system of a suitable size (which, after reading this book, you'll know how to do), how much energy your system delivers will depend on your roof. Specifically, its size, direction, angle and shading.

Before making any buying decision, you need to understand how your roof will affect your solar performance and, therefore, its payback. Here's what you need to know:

Roof size

Unless you own large swathes of land and want to 'ground mount' your system, the size and shape of your roof will determine how many solar panels can be installed on your home.

> **Online resource:** This tool I made will allow you to bring up a satellite photo of your roof and fit virtual solar panels to it: solarquotes.com.au/roof

You need to consider:

1. The *largest* system (in kW) you can fit on your roof.
2. Which directions your roof faces, and
3. What percentage of panels will point in which direction.

Do that now. Go to your computer, tablet or phone and fire up my panel-fitting tool. Watch the tutorial and then smother your roof with solar panels to discover the upper limit of what's possible for your home.

When you have your results, enter them here or on the worksheet:

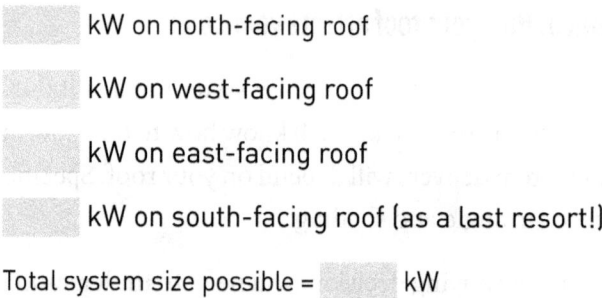

        kW on north-facing roof

        kW on west-facing roof

        kW on east-facing roof

        kW on south-facing roof (as a last resort!)

Total system size possible =   kW

Roof direction

Depending on your roof's design, you may have a choice of roof directions or be forced to use certain roof faces. At least, thanks to the tool above, you know your options. Let's see how roof direction affects your system performance and payback.

In Australia, the sun rises in the east, moves across the north and sets in the west.

North-facing roofs will capture the most sun, giving you the most solar energy over any 12-month period.

Given the choice, most installers will recommend north-facing panels. But, increasingly, people are putting some or all of their panels on east- and west-facing roofs too. A panel facing east or west will produce about 12% less power than if it faced north – so what's their logic?

STEP 1: Essential Knowledge

Figure 1.11 The path of the sun in Australia.

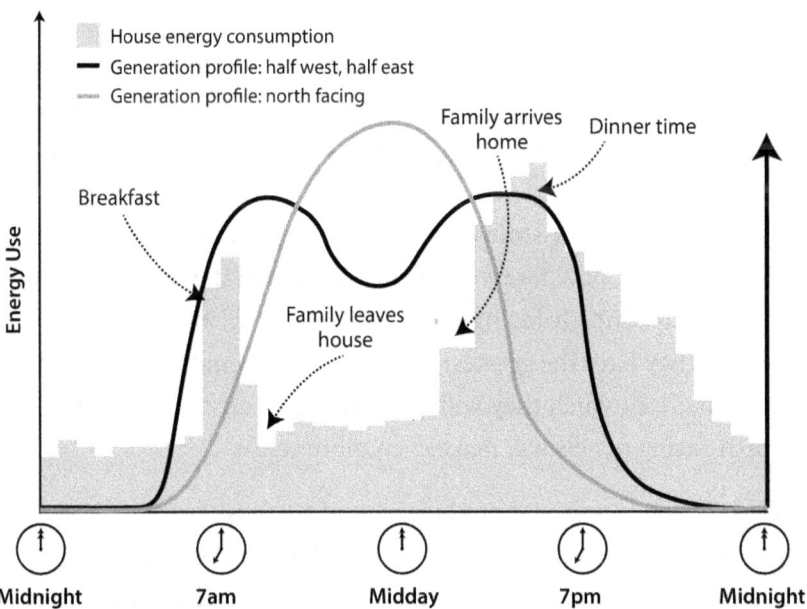

Figure 1.12 A typical home's energy consumption profile over a working day, compared to solar generation from both a north-facing system and an east/west split system.

They do it because panels that face east or west will shift some solar generation from midday to the start or end of the day. Most homes use more electricity early and later in the day, as the family wakes up and after the family gets home from school or work.

For a majority of Aussie homes, having panels facing east and west will make their solar production more in sync with their usage habits.

This means the householders will consume more of their solar electricity in their home and export less to the grid.

Which panel direction will maximise self-consumption for you? The best panel direction to maximise self-consumption of solar electricity will depend on your household's consumption patterns. Panel directions and the type of households they usually best suit are set out below.

North: Panels that face north will produce the most electricity overall. North-facing panels are often the best choice for people who are at home during the day. This is not only because they are there to use the electricity when it is produced but also because it is easy for them to shift demand by using washing machines, clothes dryers, pool filters and other devices in the middle of the day. Because north-facing panels produce the most electricity overall, they have the greatest environmental benefit. And if feed-in tariffs rise (which they will if wholesale electricity prices rise), north-facing panels will make even more sense.

West: Panels facing west produce around 12% less electricity overall than north-facing panels. They also produce less electricity in the morning but more in the afternoon. They reach their maximum output around 1:30 pm, and just before sunset they will produce around one-quarter of their peak maximum. West can be a good direction for people with high demand for air conditioning in summer. It's also an excellent direction for people who are usually out of the house by the time the sun comes up but return in the afternoon. However, because of the reduction in the total amount of electricity generated compared with north-facing panels, the increase in self-consumption that will result will need to be considerable to make west-facing panels worthwhile.

North-west: Panels that face north-west will produce around 5% less electricity overall than north-facing panels. Their electricity production through the day will be between that of north-facing panels and that of west-facing panels. They produce slightly more electricity during the afternoon and slightly less in the morning.

East: Panels that face east are similar to west-facing panels, producing around 12% less electricity overall than north-facing ones, but they will produce more electricity in the morning and less in the afternoon. People who are out of the house in the afternoon can benefit from them, as can households with high consumption in the morning.

North-east: Panels facing north-east will produce around 5% less electricity than north-facing panels. Their production throughout the day will be between that of north-facing panels and that of east-facing ones.

East and west: Placing some panels facing east and some facing west will result in the total amount of electricity produced being around 12% less than if all the panels were facing north. This arrangement is often called an 'east/west split' and has the advantage of producing a more constant output of electricity during the day (see the previous figure), which can help to increase self-consumption. The steeper the roof, the smoother the output of the system will be.

An east/west split can normally have a different number of panels facing in each direction. If a household uses more electricity in the afternoon, more panels can be placed facing west. It can suit people who are at home all day as well as those who are at home in the morning or in the afternoon.

South: For most Australians, south is the worst direction panels can face. Some people consider installing panels facing south almost as large a mistake as installing them upside down. In Sydney, south-facing panels will produce around 28% less energy than north-facing panels.

In the far north, the difference is not so great. In Townsville, south-facing panels will only produce around 17% less electricity in total than north-facing ones. In summer, they will also produce more electricity than north-facing ones. Because people in Townsville usually use considerably more electricity in summer than in winter to run their air conditioners, south-facing panels can improve self-consumption there.

In Darwin, south-facing panels only produce about 15% less electricity overall than north-facing panels. Because the Northern Territory still has a high feed-in tariff, though, the most cost-effective direction to face the panels is north.

Combining directions: Panels can be placed in multiple directions, not just an east/west split. For example, some panels could be placed facing north and some facing west. This will result in an output similar to all panels facing north-west. It's even possible to have panels facing in more than two directions.

Online resource: Power loss table:
solarquotes.com.au/powertable

STEP 1: Essential Knowledge

This table shows you how much energy you can expect to get from almost any roof direction. Bear in mind that a good installer will do all these calculations for you. They'll even project which roof directions will give you the best payback based on the compromise between energy yield and maximising self-consumption.

Roof angle

Generally speaking (unless your roof is flat), the pitch of the roof on your home is going to be the angle that your solar panels are mounted at. In Australia, common roof pitches are 15 degrees or 22.5 degrees from horizontal, so your panels will most likely be mounted at one of those angles.

If you have a standard roof at one of those angles then that is the angle your panels will be installed at. It is simply not worth the expense to engineer a framing system that will adjust the angle of the panels by a few degrees for almost no increase in energy yield. If this is you, you can skip to the section on shade. However, if you have a flat roof – or you're designing a house from scratch – read on.

Flat roofs

If your roof is totally flat, I strongly recommend mounting the panels at an angle of at least 10 degrees. This is really important because it will allow any rain to run off the panels. If the rainwater pools on the surface of a flat solar panel, it is more likely to eventually get through the panels' seals and into the solar cells themselves. If this happens, it's game over for the panel. You'll get an earth fault and you'll need to replace the panel. Many panels have a warranty condition that they must be mounted at least 10 degrees from horizontal.

Don't listen to anyone who tells you to have the panels horizontal and just add a panel or two to make up for lost efficiency. Horizontal panels are asking for trouble down the line. They'll also get a lot dirtier because they'll have little ability to self-clean in the rain, so you'll need to clean them manually more often.

For these reasons, if you have a flat roof, most good installers will add the cost of a 'tilt frame' to their quote. If you do have to have flat panels then use good frameless ones, as that will help any water flow off them and reduce the amount of grime that builds up. Frameless panels are more expensive, though, and they need to be really well sealed at the edges.

> **Online resource:** If you're mounting your panels on a frame or you're lucky enough to be designing your own solar home, then your choice of angle is less limited. So what angle should you choose?
>
> As this situation is fairly uncommon, I explain your choices online here: solarquotes.com.au/angle

Shade

This is a big one. Nothing destroys the efficiency of a solar system like shade – not roof pitch, not roof direction, not clouds. When it comes to solar power, shade is your enemy.

If your roof is mostly in the shade from 10am to 3pm, and you're not willing or it's not possible to remove the trees or other objects casting shadows, I'm afraid you're out of luck. And if an installer tells you any different, you, my friend, are in the presence of a cowboy – a con artist, a dodgy installer, the Lord Mayor of Shonkytown.

But what if you have shade that is not 'substantial'?

STEP 1: Essential Knowledge

The first thing you need to do is quantify the problem, so you can answer the question: 'How much will my shade affect the solar output on my roof?'

You will need to find an installer with a device that quantifies shade. My favourite one is called a 'SunEye'.

These shoebox-sized devices can be plonked on your roof and, within a few seconds, they'll take a 360-degree photo capturing anything that may cast a shadow on your roof. But that's not all! They then use their GPS to work out where you are on the globe, and calculate how much shade your roof will be in every hour of the year based on the sun's position above your roof over time.

Pretty smart, huh? The result is a percentage that tells you how much the shade will affect your solar production, where 0% means there is no effect, and 100% means there is zero power (for people who live in caves).

Once you have that number to rely on, you can decide if the panels will generate enough power to be worth buying.

I simply don't understand how anyone can quote a system for a shaded roof without the numbers from a SunEye or equivalent device. How can you possibly be expected to make a decision based on a wild-assed guess from an installer or salesperson without the necessary gear?

If you have shade on your roof and if you don't see such a device in your installer's hand, or you don't get a report based on the readings from a SunEye, or your installer won't guarantee the findings in the report... find an installer who will. And please don't believe the BS that they can work it out from a Google Maps picture of your roof. They can't.

For those of you with shade, you could consider a special design of solar system called panel-level optimisation, which is more tolerant of partial shading. We'll cover that in Step 5.

Have a think about your roof. Do a quick estimate to find out how many panels will fit using my online tool. While you're there, see what directions of your roof faces are available for panels, and understand the pros and cons of putting panels on some or all of those roof areas. Then look for roof features that may cast shade on your panels, such as chimneys, flues, TV aerials and plumbing vents. Finally, look for trees and other buildings that could cast shade on these roof areas. If shade is an issue, be prepared to ask any quoting installers for a shade assessment. (Or bite the bullet and pay a couple of hundred dollars for an assessment – many installers will give you this as a credit on any future purchase.)

You can now have an informed conversation with a solar installer and salesperson about the solar options for your particular roof.

Fundamental 6: The high cost of cheap solar

You bought this book because you want to make a smart investment in solar. The first part of buying smart is not to pay too much. The second part of buying smart is to not pay so little that you get a system where corners have been cut and it doesn't do the job.

I'm defining 'do the job' as: generate lots of reliable energy that will give you low bills for decades.

The main principle to bear in mind when buying solar is that the best deal is not the cheapest deal. In Australia, there are large

STEP 1: Essential Knowledge

companies operating at the bottom of the market, whose only 'unique selling point' is that they're cheaper than everyone else. With solar, there are simple ways to be cheaper than everyone else:

- Source the cheapest solar panels
- Source the cheapest inverters
- Pay peanuts for the install – ideally with low-paid full-time employees, or in busy times with the cheapest subcontractors you can find

Really cheap panels are highly unlikely to last more than three to five years. Ditto inverters. A really cheap and rushed installation will look ugly, underperform, fail prematurely, be dangerous – or all four.

To keep their marketing costs low, these companies need to maximise the effectiveness of their marketing. They can achieve this through advertising that

- promises unachievable returns through misleading assumptions about inflation of electricity prices and self-consumption
- claims their cheap panels will outperform the best on the market because of some 'secret sauce'
- claims their cheap inverter will outlast and outperform any other inverter using obscure technical jargon like 'start-up voltage' to sound impressive
- claims their panels and inverters are 'engineered' in a country that's highly respected in the industry (often Germany), and often using a PO box from that country as the research and design headquarters

Please don't fall for the allure of one of these cheap systems. Here's how to avoid this junk:

- Check your panel is one of the low-risk brands listed in Step 5.
- Check your inverter is one of the low-risk brands listed in Step 5.
- Do a quick back-of-the-envelope check of their payback claims (Step 4). If they assume 100% self-consumption, go elsewhere.
- Look at the length of their installation warranty – insist on at least five years. Use a firm that has been going for at least that long to avoid a company that has no intention of being around to fulfil its warranty. Read online reviews of these longstanding companies to see how their installations have fared.

How to make sure you don't pay too much – or too little

I maintain a web page that shows up-to-date, ballpark pricing for good-quality solar:

> **Online resource:** Up-to-date price ranges for good-quality solar: solarquotes.com.au/cost

The first thing I recommend is to check that the quoted prices are in this ballpark.

There are two legitimate reasons you could pay less and still get a reliable system:

1. Using a budget inverter – expect to pay up to $600 less per 5 kW than you would for a premium inverter.
2. Using a high-volume installer that gets a particularly good price on legitimate panels – expect to save up to $400 per 5 kW.

STEP 1: Essential Knowledge

If you live in a metro area, you may pay up to $1,000 less than the ballparks I've provided online for a reasonable budget system. But to get there, the company is going to be dependent on installing consistently high volumes to cover their overheads and it will be operating on slim margins. This increases the risk that the company will not be around in the future.

There are three legitimate reasons you could pay more than the standard:

1. You have a particularly difficult install (you need scaffolding, panel tilt frames, modifications to your grid connection, or switchboard upgrades).
2. You live way out in the bush, so they need to add travel costs.
3. You are also getting lots of extras: power-diversion hardware, battery-compatible hybrid inverters, etc.

In my opinion, the best way to protect yourself against paying too much or too little is simply to get three quotes from well-regarded installers. In Step 6, I'll walk you through that process.

Fundamental 7: Don't get screwed over, know your rights

A solar system is unusual in that it comes with a long warranty – up to thirty years. If you're buying a solar system, it's important that you understand:

1. your rights under Australian Consumer Law, which override anything else, including any written warranties, and
2. the specific details of warranties that come with solar systems.

As a consumer living in Australia, you're lucky to be protected by some of the strongest consumer laws on the planet. You have rights under Australian Consumer Law that no one can override. If I were to sum up the spirit of Australian Consumer Law it would be: if it sounds like a vendor is being unreasonable, they're probably breaking Australian Consumer Law.

Most people have no idea of their rights under Australian Consumer Law. Most people think that they're beholden to the vendor's or manufacturer's written warranty. Nope.

Here's a simple example from the mobile phone industry. Imagine you pay $1,200 for a new smartphone with a one-year warranty. After 13 months, the battery dies. You are out of warranty, right? Probably not. If a typical person would consider the failure to be unreasonable based on, among other things, the price paid for the product, Australian Consumer Law says the vendor should bloody well fix it. And I would say that any reasonable person would expect a $1,200 phone to last more than 13 months. If it were a $50 Aldi special, you'd have a much tougher argument.

Here's another example. You have a solar system and organise a five-yearly inspection by a solar electrician who is not from the original solar company. Six months later, the system fails and the solar company says the warranty is void because you used someone other than them for the inspection. That won't hold up, because it's unreasonable. You can use any qualified person you like to maintain or even repair a product and it cannot legally void your warranty. As I write, Apple are likely to be walloped by the Australian Competition and Consumer Commission (ACCC) for refusing to repair iPhones under warranty that had a third party replace a smashed screen. If a company like Apple can get

STEP 1: Essential Knowledge

Australian Consumer Law wrong, I'm afraid even the good installers can.

Just remember: if it sounds like a vendor is being unreasonable, they're probably breaking Australian Consumer Law. And if they refuse to comply, you can take them to your local court or tribunal where you may be pleasantly surprised at how pro-consumer our laws are.

Usually, you just have to threaten to go to tribunal and they'll capitulate if they've done the wrong thing.

Solar system warranties

Solar systems come with four or five main warranties:

- Panel performance warranty – 25 years (industry standard)
- Panel product warranty – typically 10 to 15 years
- Inverter warranty – usually 5 years (companies sometimes offer this with an optional upgrade of 10 years or more, depending on the manufacturers)
- Battery warranty (if you choose to add batteries)
- Installation warranty – don't accept less than 5 years. This can be up to 10 years in some cases (provided by the installer)

Under Australian Consumer Law, all five warranties are the absolute responsibility of the entity that you gave the money to and entered into a contract with: the solar company. If, for example, your panels fail and the solar company gives you the number of the panel manufacturer and tells you to make a claim directly, they can't do that. They need to manage the warranty claim with the manufacturer, fix everything up for you, and then claim any money they are owed from the manufacturer. If a hardware manufacturer fails to honour their obligations, the solar company has to wear it – not you.

If you ever find yourself in the unfortunate situation where you need to make a warranty claim but the solar company is no longer in business, you can make a direct claim with the company that imported or manufactured the hardware. For this reason, it is important to insist that the handover documentation contains all the hardware manufacturer's Australian contact details. I'll go over what should be in this documentation in the final step.

Also bear in mind that if you use a solar company that imports their own hardware, there is only one company that you can go to for a warranty claim. There is no backup company if they stop trading.

> **Online resource:** This blog post describes seven other scenarios where Australian Consumer Law can give you rights that you (and many solar companies) are probably unaware of: solarquotes.com.au/acl

Summary

- Kilowatts (kW) measure power. Power is how quickly you are generating or using energy.
- Kilowatt-hours (kWh) measure energy. Energy is how much power you have stored, used or generated over time.
- It is important to understand the difference between a kW and a kWh, or solar and batteries will quickly get confusing.

STEP 1: Essential Knowledge

- The solar rebate is the discount on the price of a solar system at the point of sale. All prices you see advertised will already include this, and it usually reduces the cost to buy solar by a third.
- The feed-in tariff is the amount you get paid for exporting your excess solar into the grid. It is generally 6c to 16c per kWh depending on competition where you live.
- Solar systems can be on grid, off grid or hybrid. If you want to save money and you already have a grid connection, on-grid solar is the rational choice.
- Self-consumption refers to how much of the solar energy you generate gets used by your home. The more solar you self-consume, the better the economics of solar will be for you.
- When you get solar you need a 'net meter'. This will measure how much solar is exported, so you can be paid for it, and how much grid electricity is imported, so you can be charged for it. The net meter does not know how much solar is generated or how much energy your home is using. To measure that you'll need extra equipment.
- The best direction to place your solar panels is on a north-facing roof, if you want to get as much energy as possible over a year. It is also fine to place them on east- or west-facing roofs, where they will give you more energy in the morning and late afternoon respectively, but will produce about 15% less energy overall.
- The best angle to place your panels at is the angle that your roof is already built at. If you have a flat roof, angle the panels at least 10 degrees to allow them to self-clean in the rain.

- Don't buy the cheapest systems on the market; they are unlikely to last long. Pay a fair price to a good installer who is likely to still be around in a few years.
- Your panels will come with a warranty of at least 25 years, so you need to understand your rights in case anything goes wrong. Australian Consumer Law is pro-consumer and can override written terms and conditions. If a company acts unreasonably, the chances are that they are violating Australian Consumer Law.

STEP 2: Measuring Your Energy Use

> **WARNING:** This step involves opening your switchboard to look inside. Your switchboard contains deadly voltages. Do not touch anything in the switchboard. If you are not comfortable opening your switchboard, don't do it.

Now you have a grounding in the basic knowledge needed to understand solar and how it relates to your home and bill, you're almost ready to jump into details of which solar panel brand to buy and how many of them to whack on your roof.

Before we do that, we need to step back and think about the problem we are trying to solve with those panels.

95% of people who buy solar tell me their main motivation is to lock in low bills for decades to come. But how can you be sure what your new bills after going solar will be?

The main factor that determines the payback of a solar system on your home – and, therefore, determines what your new bills will be – is your 'self-consumption ratio'.

Your self-consumption ratio is the percentage of generated solar electricity that you use in your house (the rest being exported to the grid).

If you can accurately predict how much of your solar energy will be self-consumed, you can accurately predict your new bills. If you can accurately predict your new bills, you can make an informed decision about whether to buy solar, how to finance it and what size to buy, based on your personal criteria for payback.

Accurately predicting your self-consumption takes some effort. This step describes how to make that effort.

> **Tip**
>
> If you don't want to do all the work outlined in this chapter – calculating your actual self-consumption ratio – the good news is that you don't have to!
>
> The lazy or impatient among you can assume that your self-consumption ratio will fall into a range:
>
> - Assume your worst-case self-consumption is 10%.
> - Assume your best-case self-consumption is 70%.
>
> In Step 4, work out your worst- and best-case savings range using the two values above.
>
> If you're happy with worst-case savings, then you can get solar safe in the knowledge that you'll get savings in the range you've calculated.

For those of you who want to predict your solar savings more accurately, here's how to perform a quick and dirty energy audit that will reveal your self-consumption ratio.

To get there we need to measure how much of your energy you use during the day – when the sun is shining – and how much you use at night. With these numbers, you can make a good estimation of solar's financial return on your home before you get quotes (Step 4). You can also use your measurements to confirm the financial projections when you get firm quotes (Step 6).

STEP 2: Measuring Your Energy Use

Energy audit step 1: Benchmark your gross daily usage

You can easily find out how much energy you use over 24 hours – on average. It's printed on your electricity bill.

Figure 2.1 is an example from an Origin energy bill:

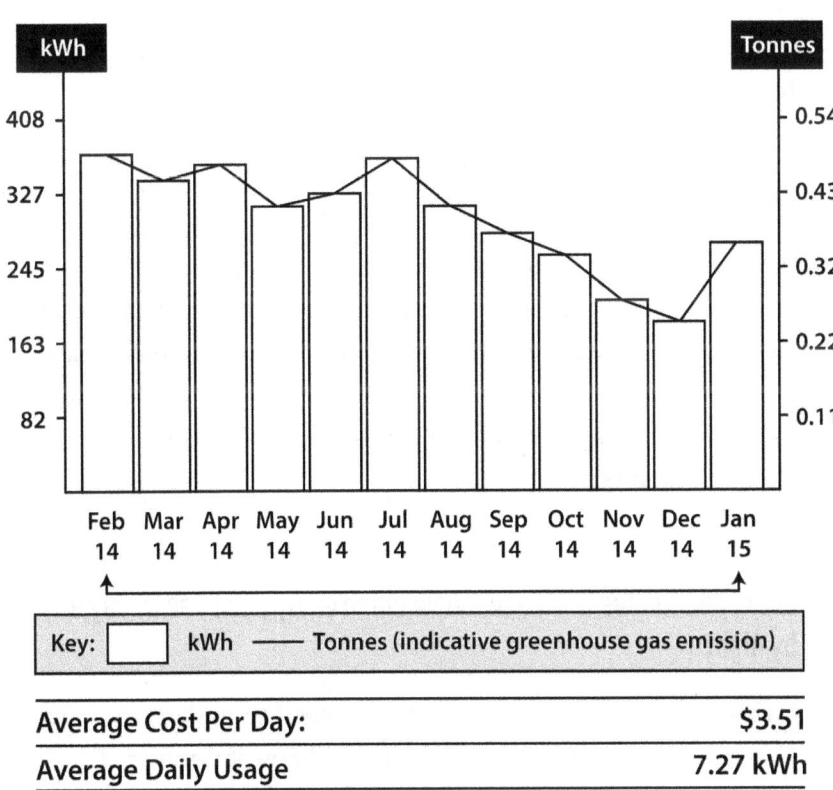

Figure 2.1 How a typical electricity bill presents your daily consumption.

This household uses 7.27 kWh per day, averaged over the year (a very efficient home).

Dig out a recent bill and find your average daily usage. It's probably between 7 kWh and 30 kWh, depending on your lifestyle and home efficiency.

Record it here or on the worksheet (solarquotes.com.au/worksheet):

> My annual daily grid usage (from my bill): _____ kWh

(You can come back to this number after you've got solar to see how it's fallen.)

Now we have the absolute minimum we need to get started. We know your average grid usage for 24 hours.

Bear in mind that this figure may vary with the seasons. If you have electric heating or cooling and you use it a lot more in summer and winter than in spring and autumn, your daily usage for the winter and summer will probably be around 6 kWh more. But let's not complicate things – in my experience, the average values are good enough.

Energy audit step 2: Measure how much energy you use by day and by night

To measure your energy use, you first need to understand how electricity is connected to your house and how it is metered.

From answering thousands of questions about Australians' electricity use, I've learned that many householders:

STEP 2: Measuring Your Energy Use

- don't know what kind of electricity connection they have (single phase, two phase or three phase)
- don't know what kind of tariff they are signed up for (flat rate, time-of-use, demand tariff, economy tariff or controlled load tariff)
- can't interpret the individual line items on their bill, or
- don't know how to read their meter

This lack of understanding is so widespread because metering and billing is confusing.

If you want to take control, you'll need to put some work in to understand your electricity connection, your metering and your bill.

Your electricity connection

Let's start with how your home is physically connected to the grid. It will be:

- single phase
- two phase, or
- three phase

If you have two phase or three phase, each phase may have a separate meter or they may all go into a single meter.

Single phase, two phase and three phase explained

Single phase. Most Australian homes have 'single phase' electricity. This means that they have one live wire to their home carrying all their electricity. This wire is called the 'active' conductor. There is a second wire, called the 'neutral conductor', which provides a return path because – as you should remember from primary school science – electricity always needs a circuit to flow round.

Two phase. Two phase supplies are rare, but some people have them. Two phase means there are two live wires going into your home.

Three phase. As people get bigger homes and more powerful appliances, such as large air conditioners and pool heaters, they may need more power than can flow down a single wire. For this reason, more and more homes are getting 'three phase' connections. As you have probably already guessed, three phase means that you get three live (or active) wires instead of one. It follows that you can have three times the power coming into your home.

If you have three phase power, the phases are called red, white and blue. Your single phase appliances run off one of those phases. Any large, three phase appliances are connected into all three phases.

As electric cars become common, people will start to upgrade to three phase so they can charge the car more quickly.

The first thing we want to determine is: how many phases do you have?

If you know how many phases you have (one, two or three), then record it here or on the worksheet:

> My home has a ____ phase supply.

If you aren't sure how many phases you have, you could call your electricity retailer and ask. If one of your neighbours is technical, you can ask them.

Or, you can snap a picture of your switchboard and email it to: support@solarquotes.com.au.

STEP 2: Measuring Your Energy Use

Make the subject of your email 'How many phases do I have?' and we'll work it out for you.

Do you have a 'controlled load' supply?

Now you know how many phases you have, we need to see if you have a special separately metered, cheaper tariff for running your electric hot water (and perhaps your underfloor heating or pool pump too). This tariff is called:

- off-peak
- controlled load
- economy, or
- dedicated circuit consumption

It could also have some other obscure name, depending on your retailer and local network. This cheaper tariff may need a separate meter, or it may be measured inside one of your other meters.

If you are in Western Australia, you won't have one of these tariffs because they don't do them there. In the rest of Australia, you may have one if you have electric hot water, underfloor heating or a pool pump. The best way to find out is by looking at your bill.

Most bills will list all your tariffs on the back page, as in Figure 2.2.

This example is for a poor bugger with a very complicated electricity tariff. She is on a time-of-use tariff, which charges different amounts depending on the time of day.

The Good Solar Guide

Electricity charges: 11 May 2014 – 10 August 2014 (92 days) ?	Rate $ per kWh	Total $
Peak – first 411 kWh	$0.3124	$128.40
Peak – next 452 kWh	$0.3245	$146.67
Peak – next 389 kWh	$0.3455	$134.40
Off peak – 628 kWh	$0.1250	$78.50
Shoulder 1 – 455 kWh	$0.2750	$125.13
Shoulder 2 – 256 kWh	$0.2365	$60.54
Controlled load – 408 kWh	$0.1050	$42.84
Service to property charge – 92 days	$0.5384	$49.53
Total electricity charges (Ex. GST)		**$766.01**
Total electricity charges (Inc. GST)		**$842.61**

Figure 2.2 Range of tariffs listed on the back of an electricity bill; most people only have one or two.

She also has a controlled load tariff, which is the cheapest tariff of the lot at 10.5c per kWh.

If you're on a time-of-use tariff like this one, please consider moving to a simple standard tariff.

> **Online resource:** Why time-of-use tariffs are generally bad news: solarquotes.com.au/tou

If you're on a simple flat tariff like this one, the back of your bill may look like Figure 2.3:

STEP 2: Measuring Your Energy Use

New charges and credits

Usage and supply charges	Units	Price	Amount
Peak	419 kWh	$0.2615	$109.57
Tariff 33 Controlled load	331 kWh	$0.1898	$62.82
Supply charge	96 days	$0.8581	$82.38
Total charges			**+ $254.77**

Figure 2.3 A bill with only 'peak' and 'controlled load' tariffs.

Here, the controlled load is obvious at 18.98c per kWh.

Write down if you have a controlled load tariff here or on the worksheet:

> I have a controlled load tariff: (Yes/No)

If you can't work it out from your bill, send a copy of it to support@solarquotes.com.au with the subject 'controlled load?' and we'll decipher it for you.

Getting your consumption numbers from your meters

Now you know how many phases you have, and whether you have a controlled load tariff, we can get the numbers we need from your meter or meters.

If you have a smart meter that transmits its data wirelessly to your retailer, you can ask them for the data. It is logged every 30 minutes, so it's perfect. The downside is that it can take many weeks for the retailer to log into their system and press the big red button that says 'send customer their data'. If you do have a smart meter, ring up your retailer and ask them for this data now. You might be lucky and get it quickly, or you may be able to log in and see it immediately.

If you don't want to get your energy retailer involved, don't want to wait, or, like most Aussies, don't have a smart meter, traditionally you'd have to spend a couple of hundred dollars on an energy monitor and another couple of hundred to get it installed by an electrician.

We're going to do this the cheap and cheerful way – manually. You're simply going to have to look at your meter (or meters) four times.

We want to break our meter readings into two chunks a day, giving us a night-time usage and a daytime usage. We'll do this by taking meter readings at 7am and 7pm on a work day and a non-work day. If you don't work, you should still take readings on a weekday and on the weekend – as you probably still behave differently on a weekend.

It is important that you do this on a typical usage day. If you don't normally have your heating or cooling on, don't choose a day when the air conditioning is cranking.

How to read your meter

First, count your meters.

Have a look in your meter box and count how many electricity meters you have.

The meter is a box about the size of a brick. It will have either a digital or a mechanical numeric display.

If you have one meter, that's the one you need to read.

STEP 2: Measuring Your Energy Use

Got two meters? Then you have:

- single phase power and one controlled load tariff (most likely), or
- three phase power and one controlled load tariff (next most likely), or
- two phase power and no controlled load tariff (rare), or
- two phase power and one controlled load tariff (rare)

If you aren't sure, send a picture of your meter box and a copy of your bill to support@solarquotes.com.au and we'll advise what meters you've got and how to read them.

If you have two meters, the most likely configuration is that one is for your regular tariff and one is for your controlled load tariff. If they are not labelled, put the oven and kettle on and see which meter goes up the quickest. This is your 'regular-tariff' meter (the expensive one), and the one we want to read for this exercise.

Economy/controlled load tariffs explained

Most people in Australia (especially those in New South Wales, Queensland and South Australia) can use electricity at a cheaper rate to power their water heater (and sometimes their pool pumps). This is commonly referred to as 'off-peak hot water', although the retailers call it a 'controlled load' or 'economy' tariff.

It is called a controlled load because the electricity company can choose when to supply the electricity to your hot water system. They literally control your load. This electricity is cheaper because they only supply electricity on this tariff when wholesale energy is cheap.

To control your load and charge you a separate rate, they either fit a second meter for your controlled loads or have a single multi-function meter with two outputs and two digital displays – one for your regular loads and one for controlled loads.

Why we don't read the controlled load meter in this exercise

If you get solar, the output of the system is usually wired into the 'regular' meter input[5] so your solar offsets the most expensive electricity.

For this reason, we only need to read your regular meter to work out how much solar will save you. Your solar will not be connected to your economy meter, so it will make no difference to that part of your bill. This isn't a big deal, as the controlled load (see Figure 2.4) is usually a small fraction of your total bill, and if you follow the advice in **STEP 3: Heating Your Water**, it will reduce or be eliminated altogether.

Got three meters? Then you almost certainly have three phase power where each phase is metered separately. This style of metering is slowly being replaced with a single meter that does the same job. But because you still have the old-school three-meter system you'll need to read all three meters and add them up to get one total figure. If you get solar, these three meters will be replaced by a single meter that does everything.

---

[5] Some people who have a high consumption on their controlled load choose to add a second solar system and wire it into their controlled load meter, but this is rare so I won't cover it here.

STEP 2: Measuring Your Energy Use

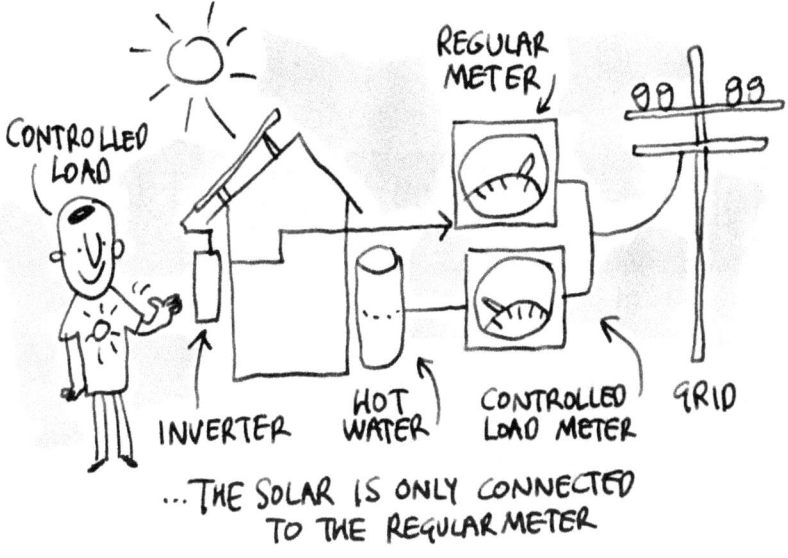

Figure 2.4 The solar is only connected to your regular-tariff meter; the controlled load meter remains separate.

Got more than three meters? This is getting silly: send a picture of your switchboard and a copy of your bill to support@solarquotes.com.au and we'll work out what's going on.

First, I'll explain how to decipher the meter's dials or display; then, I'll explain when to read it.

How to read a mechanical meter with dials

If you have a really old meter, it will have dials that you have to read. These meters are rare, and I give instructions on how to read them here: solarquotes.com.au/oldmeter.

How to read a mechanical meter with reels

Figure 2.5 A mechanical meter with reels.

The meter shown in Figure 2.5 is super easy to read – the only thing to look out for is the decimal point (shown as a comma here).

This meter reads 36,879.3 kWh (it is actually halfway between .3 and .4 – but a tenth of a kWh is nothing in the scheme of things, so don't stress out about the last digit. The one that's most visible is the one to go for. In this case, either 3 or 4 will do).

How to read an electronic meter

An electronic meter will have a digital display. It will usually switch between a few different readings, displaying each one for about five seconds.

If you don't have a controlled load tariff or time-of-use tariff then you need to wait for the display to show ' kWh'. There may be a large space between the ' kW' and the 'h'. Then write down the numbers that are displayed largest.

For example, this meter's reading is 16,266.5 kWh, and it is shown on the screen labelled '007'. (A Licence to Bill?)

STEP 2: Measuring Your Energy Use

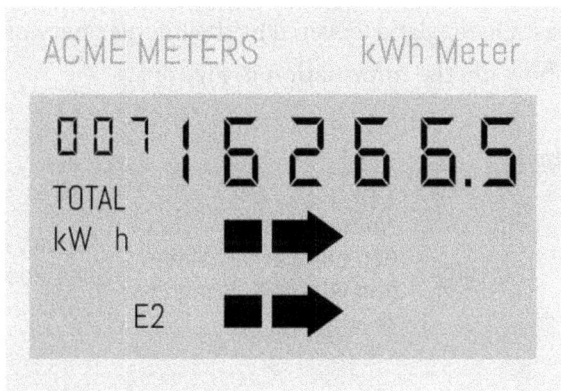

Figure 2.6 A digital meter.

If you have a controlled load tariff and only one meter, or time-of-use billing, then one of the kWh displays will show your total non-controlled usage and one will show your controlled load usage. You may also have up to three other displays showing your use at off-peak, peak and shoulder times. Confusing or what?

The best way to ensure you read the correct number is to type the following text into Google:

'[your electricity retailer] how to read my meter'

This should give you a page that shows what screen you need to read to get your 'total energy' or 'peak energy' usage. The information is often based on who your local electricity network is.

> **Online resource:** Your energy network (called a DNSP – distributed network service provider) is different from your electricity retailer. This link will show you who your local energy network is: solarquotes.com.au/dnsp

For example, if I got my bills from Simply Energy and lived in south-east Queensland (covered by the Energex network), Google would show me the information in Figure 2.7.

Queensland

Energex	EM1000	No registration number (blank) (total energy)	Minus sign (-)	A (peak energy), B (shoulder energy), C (off-peak energy)
	EM1200 and EDMI Atlas	01 (total energy)	40	05 (peak energy), 10 (shoulder energy), 20 (off-peak energy), 30 (second circuit such as hot water or slab Heating)

Figure 2.7 Meter-reading information from Simply Energy's website.

I want to know 'total energy'. This information is in the third column. If my meter says EM1000 on it, I read the screen with a 'blank' label. If my meter says EM1200 or EDMI Atlas, I read screen '01'.

They don't make it easy, do they? The system is designed to confuse, so if you want to skip this step, you can, as described earlier.

Collecting your meter data

Now you know how to get the numbers, whether you have a mechanical or an electronic meter, I'm going to show you how to quickly, and without cost, measure your daytime usage for a typical work day and a typical non-work day in your home. This is because

STEP 2: Measuring Your Energy Use

being at home makes a big difference to how much energy you use (who'd have thought!).

We'll also measure typical night-time use, but we won't differentiate between work nights and non-work nights, because in my experience they don't differ much. After all, you're asleep for most of the time.

We'll do this by reading your meter four times in 48 hours and then doing some sums with the results.

If you have a smart meter and an online portal, you don't need to do the manual readings. You can log in and get these numbers instead. For the rest of us, read on.

How to measure your two daytime usages and typical night-time usage

I'll walk you through the process for Sunday (non-work daytime) and Monday (work daytime). Of course, if you aren't a typical worker bee, feel free to use any two days of the week that capture your work and non-work lifestyle the best.

> **Online resource:** The most difficult thing about this exercise is remembering to take the readings. To make it easier to remember, I've created a calendar file that will prompt you to take the four readings with four alarms. Simply go to this link on your smartphone: solarquotes.com.au/alarms

We're going to read your meter at the following times:

- Sunday 9am
- Sunday 7pm
- Monday 9am
- Monday 7pm

Then we'll put the readings in a table like this:

	9am	7pm
Sunday		
Monday		

There's a blank, printable copy of the data-collection table on your worksheet: solarquotes.com.au/worksheet.

Example

Here's the data-collection table printed out and completed for my house:

	9am	7pm
Sunday	16030·4	16038·6
Monday	16042·2	16051·3

I then put the four numbers into this online calculator: solarquotes.com.au/calc.

It pops out my work and non-work daytime usages, and one night-time usage. Figure 2.8 shows how the simple calculation works. The figures are typically all between 3 and 30 kWh. If yours are not, double-check your readings.

My daytime usage – Sunday: 8.2 kWh
My daytime usage – Monday: 9.1 kWh
My night-time usage – 3.6 kWh

It also spits out my daytime usage averaged over the whole week, based on a week containing 5 work days and 2 non-work days: 8.46 kWh.

STEP 2: Measuring Your Energy Use

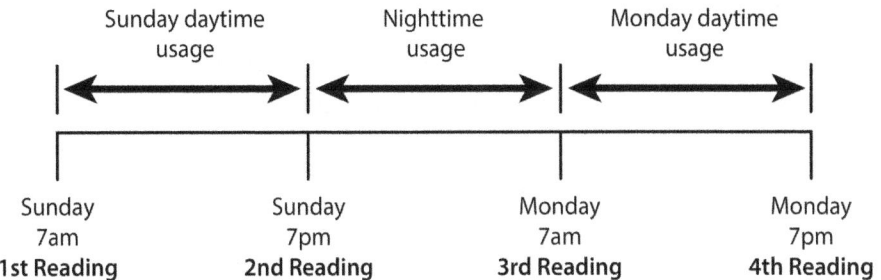

Figure 2.8 Usage readings

The weekly average daytime usage and the night-time usage are the two magic numbers that we need to:

1. Get a preliminary size for your solar system, and
2. Estimate the financial returns of the system.

Now's the time to make plans to record your meter four times this weekend (or pull the numbers from your smart meter portal). Once you have the readings, crunch the numbers online.

Record the results on your worksheet:

Weekly average daytime usage: _____ kWh

Night-time usage: _____ kWh

We'll use these numbers to work out your personal solar payback in Step 4. First, we need to have a good look at how you heat your hot water. You have a choice to make that will affect how many panels you need and how fast they'll pay you back.

Summary

To predict potential savings from solar accurately you need to estimate your solar self-consumption. You have two options:

- If you want to move through the seven steps with no delay, simply use a worst- and best-case estimate for your self-consumption ratio. I suggest a worst case of 10% and a best case of 70%.
- If you can commit to understanding how your house is metered, then take four meter readings over next Sunday and Monday. You will be rewarded with a much better estimate of your self-consumption ratio, which will lead to a more accurate solar savings estimate in Step 4.

STEP 3: Heating Your Water

By now, you should:

- be clued up on the fundamental solar knowledge necessary to be informed when talking to a solar installer, and
- have guessed, or ideally measured, how much electricity you use in in the day, and how much at night

This means you're almost ready to decide how big your solar system should be and work out how quickly it will pay for itself, so you can decide whether to invest in solar or not.

There's one more thing you need to do first. You need to decide how your solar-powered home will heat its water. We have to do that now because your decision can affect how many panels you'll need.

Water heating is responsible for up to 28% of a typical energy bill – so it needs your attention as part of this project. Remember, the project is all about getting your bills right down.

Your existing system

If you already have an existing solar thermal hot water system and you are happy with it, then you can jump to Step 4. If you don't know what a solar thermal hot water system is, or currently heat your water with gas or electricity, then read on.

How your future solar home will heat its water depends on your current situation. You could:

- have an existing gas hot water system
- have an existing electric hot water system, or
- be building a home and have no existing hot water system

Scenario 1: Existing gas hot water system

If you currently have a gas hot water heater, you may think that a solar electricity system will have nothing to do with your hot water. You may even be about to skip this step entirely. But hold on a second. Let's quantify the problem by estimating how much gas you currently use to heat water in a year.

Most four- to five-person homes will spend about $1.50 a day on gas for water heating – about $550 per year. If you have a one- or two-person home, this will be closer to a dollar a day.

If you use LPG gas to heat your water, you'll probably be paying through the nose for it. LPG is about three times the price of mains gas.

If you are happy with your water-heating cost, and happy to keep your gas connection, then jump to Step 4. If you want to knock 70% to 80% off your water-heating bills I recommend ripping out the gas hot water system and replacing it with solar-electric hot water. This gets you one step closer to a simpler, cheaper-to-run all-electric house. Jump past Scenarios 2 and 3 to 'Your options for solar hot water'

STEP 3: Heating Your Water

Scenario 2: Existing electric hot water system

There are two types of electric hot water. The most efficient is the solar heat pump. Although heat pumps are a relatively cheap and very efficient solution in many parts of Australia, sadly they are installed in less than 2% of Aussie homes.

Heat pumps are pretty simple to understand: get an air conditioner, run it in heating mode, and use the heat it blows out to warm a well-insulated cylinder of water. A heat pump can be three and a half times more efficient than a conventional 'kettle-style' electric heater. For every 1 kW you put in you can get over 3 kW of heating out. Sounds like a magic pudding machine? Well, here's how it's done:

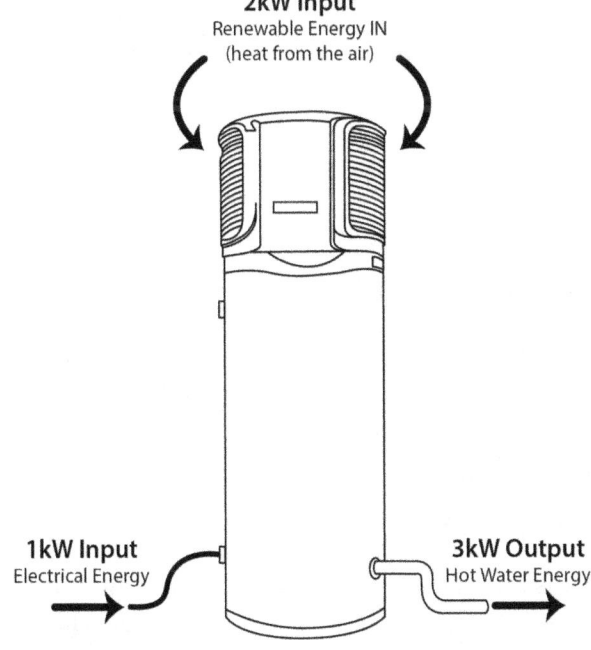

Figure 3.1 Energy flows in a heat pump hot water system.

81

If you have a heat pump and it doesn't need replacing any time soon, skip the rest of this chapter. You already have a very efficient hot water system that will soon be powered by your new solar panels.

If your water heating is electric, chances are you don't have a heat pump, as they are rare and more expensive than a regular 'resistive' hot water cylinder. A regular electric system is like a big kettle. One or two big kettle elements get really hot and heat the water.

If you've got one of these then you need to find out if it uses a different, cheaper tariff than the rest of your home. This tariff is often called a controlled load tariff, and I described it in Step 2.

If your hot water cylinder is on a much cheaper tariff (and perhaps your slab heating and pool pumps too) then, financially speaking, offsetting it with solar is a low priority. Your biggest savings will come from offsetting your regular tariff that powers everything else in your home.

If you have controlled load, you now have to make a decision. Do you leave your hot water and controlled load tariff alone, or do you go the whole hog and make your hot water solar along with the rest of your appliances?

To help you decide, let's see how much you're paying to heat your hot water with your controlled tariff. If you have a year's worth of bills, now is the time to get them out. Add up all the 'controlled load' line items for the year. That is how much you pay, per year, for hot water energy.

Figure 3.2 is an example from a quarterly bill. Add these from a year's worth of bills to see the most you could save in a year with solar hot water.

STEP 3: Heating Your Water

New charges and credits

Usage and supply charges	Units	Price	Amount
Peak	419 kWh	$0.2615	$109.57
Tariff 33 Controlled load	331 kWh	$0.1898	$62.82
Supply charge	96 days	$0.8581	$82.38
Total charges			+ $254.77

Figure 3.2 Controlled load usage on a typical bill.

If you don't have a controlled load tariff, you can estimate how much hot water energy costs you with this rule of thumb:

Two-person households use around 4 kWh a day for heating water.

Three- to four-person households use around 5 kWh.

Households of five people or more use around 6 kWh.

Reduce this by around 1 kWh a day if you live in Brisbane or Perth and by 2 kWh if you live somewhere hotter.

Multiply this by your usage tariff, then multiply again by 365.

As an example, in South Australia a four-person house will typically use 5 kWh x 365 days = 1,825 kWh. That will cost 1,825 kWh x $0.24 = $438 per year.

Write down your estimated annual hot water electricity cost here:

_____ kWh per day x 365 x _____ c per kWh = $_____ per year

Even the best grid-connected solar hot water system is unlikely to provide 100% of your hot water energy, so expect to save about 70% of that cost with solar hot water.

Now you know the numbers, the choice is yours. If you have electric water heating and want to knock 70% to 80% off your water-heating bills, read on to discover your options for solar hot water. If your hot water electricity bill is surprisingly cheap then feel free to leave your hot water alone and I'll see you in Step 4.

Scenario 3: Building a home from scratch

The third scenario is that you're building a home. I would strongly recommend designing your house to be 100% electric. You can have no initial and daily gas connection fees if you design solar hot water in from the outset. The rest of this chapter will show you your options.

Your options for solar hot water

Whether you're building a home from scratch or upgrading your gas or electric system, your new or modified hot water system can be one of four configurations:

- Solar thermal hot water
- Heat pump
- Diverted solar PV[6]

---

[6] PV stands for 'photovoltaic' and is simply a fancy name for electricity generated directly from solar panels.

STEP 3: Heating Your Water

The rest of this chapter will determine which one of these is right for you. First, we need to understand the difference between these three hot water systems.

Configuration 1: Solar thermal hot water

Did you know that there are two fundamentally different ways to generate solar energy, and, therefore, two fundamentally different types of solar panel?

The two concepts both have multiple names, but from here on I'm going to call them 'solar PV' and 'solar thermal'.

Let's start with solar PV panels. This is what I've been talking about in this book up to now, and it's what you probably imagine whenever you think about solar panels.

A solar PV panel is the name most industry professionals give to the type of solar panel that generates electricity directly from sunlight (when they're being careful with their language). If they're not being careful, they'll simply call it a solar panel and assume you know what they mean.

A solar thermal panel is a different beast. It is used *only* for heating water and installed on millions of Aussie roofs as part of a solar hot water system.

Solar thermal panels heat up the water directly, using the heat of the sun. No electricity is generated. There are two types of solar thermal panel available: flat-plate and evacuated-tube collector.

A flat-plate solar thermal panel looks similar to a solar PV panel, except it is about three times as thick. A more expensive evacuated-tube collector looks completely different – see Figure 3.3.

Figure 3.3 Flat-plate and evacuated-tube solar thermal panels.

> **Online resources:** To learn more about how a flat-plate collector works, visit this link: solarquotes.com.au/flat
>
> For information on how an evacuated-tube collector works, and when they can make sense, go here: solarquotes.com.au/tube

Both these types of solar thermal panels are designed to take cold water in, heat it with energy from the sun and send piping hot water out.

Every home in Australia needs hot water. Using fossil fuels to heat water, whether it is directly with gas or indirectly with coal-fired electricity and a big kettle, is wasteful in sun-blessed Australia.

An efficient and sustainable way to heat your water is to expose it to sunlight with a simple solar thermal panel and then store the solar-heated water in a well-insulated tank. This is a 'solar thermal hot water system'.

Am I simply going to recommend you get a solar thermal hot water system along with your solar PV system?

STEP 3: Heating Your Water

No. And the reason is economics. Although solar thermal was, for decades, the cheapest form of solar water heating in Australia, the plummeting price of solar PV means that this is no longer the case.

A solar thermal system requires:

- pipes to the roof
- a tank on the roof, or a pump from a ground-mounted tank that requires maintenance and replacing every few years
- solar thermal panels made of glass, metal and copper, and valves
- often, a crane to install, and
- a plumber and electrician on site for installation

This makes it expensive. Expect to pay from $4,000 for a flat-plate system and up to $8,000 for a more efficient evacuated-tube system.

The cheaper options for solar water heating, which can be just as efficient to run and just as environmentally sound, are the heat pump or diverted PV systems.

Configuration 2: Heat pump

A heat pump extracts heat from the air and transfers it into the water. As the heat originally came from the sun, it can be thought of as a solar heat pump.

> **Online resource:** You can read the technical details of how they work here: solarquotes.com.au/heatpump

It looks like an air conditioner piped to a hot water cylinder, as in Figure 3.4.

Figure 3.4 A non-integrated hot water heat pump.

Pretty ugly, granted. But the aesthetes among you will be pleased to know that you can also get sleek all-in-one units (Figure 3.5), with the heat pump integrated into the top of the cylinder:

Figure 3.5 An integrated hot water heat pump and cylinder.

STEP 3: Heating Your Water

If you live outside Queensland or the tropics then, give or take $100 a year in electricity costs, modern heat pumps are about as efficient as a flat-plate solar thermal hot water system but much cheaper to buy. You can claim the solar rebate on a heat pump, which reduces its cost, starting at $2,000 installed (compared with $4,000 to $8,000 for a solar thermal hot water system).

In Victoria, if you're replacing a resistive hot water heater (i.e. the big kettle) with a heat pump, you can get a heat pump for under $1,000 because the Victorian government applies a second rebate, known as a Victorian Energy Efficiency Certificate (VEEC).

> **Tip**
> In Queensland and the tropics, a conventional solar thermal hot water system will be more efficient than a heat pump – so Queensland folks should bear that in mind.

A good heat pump is as quiet as a good-quality air conditioner and will work well even in freezing temperatures, albeit with a reduced efficiency. If you don't mind the hum and are happy with paying for regular maintenance, a heat pump is a good choice. A heat pump is also a good choice if you have a generous feed-in tariff, as its efficiency leaves more electricity available for export.

If you do buy a heat pump, it's important to try to run it off solar PV as much as possible. That means two things: running it in the day (when it is more efficient anyway, due to higher outside temperatures) and having enough PV to power it over and above your other appliances. If you get a heat pump, consider going for a large (6 kW or more) PV system so that you get enough solar electricity between 11am and 3pm to run the heat pump even in winter.

No matter where you live in Australia, whether it's the tropics or Tassie, there is a potentially even cheaper option for heating water with the sun, provided you buy it at the same time as your new solar PV system.

Configuration 3: Diverted PV solar hot water system

This is the cheapest (in terms of upfront cost), most reliable and lowest-maintenance system for heating hot water from the sun. If you have the room on your roof and are concerned about the noise, cost or maintenance of a heat pump, it is a good alternative.

A diverted PV system uses an intelligent control box to divert 'spare' solar electricity from your existing solar PV panels (the ones you are about to buy) to heat your water using a cheap-as-chips conventional hot water cylinder.

This spare electricity would otherwise be exported to the grid, and, as you learned in Step 1, self-consuming solar electricity is the most valuable way to use it.

The beauty of a diverted PV system is that it will only cost about $1,000 extra to get it installed if you get it at the same time as a new solar system.

'But, Finn!' I hear you cry. 'If I'm powering my hot water as well, won't I need extra panels?'

Yes, you will. But because of the diverter's intelligent controller, which scavenges spare solar energy whenever it is available, you only need 1.5–2 kW of extra panels. And the good news is this: if you're buying a 5 kW system (one of the standard sizes sold), you can usually get an extra 1.6 kW of panels for next to nothing.

STEP 3: Heating Your Water

Sounds too good to be true? It really is true. And it's due to the way the rebate works. The rebate is based on the number of solar panels. At the time of writing, the rebate is worth more than the cost of the panels. Combine that with the fact that you can add panels above the kW rating of your inverter (see Step 5 for details) and you can usually upgrade from 5 kW to 6.6 kW for around $1,000.

> **Tip**
> When buying a new system, adding extra panels is cheap, so go big.

The combination of the solar rebate and the inverter oversizing rules can make diverted PV the cheapest upfront option for heating your hot water with solar. If you already have a conventional resistive hot water cylinder, it may only add about $1,000 to the price of your system. If you need to buy a cylinder too, budget another $1,000 to have a conventional hot water cylinder installed (I would get a dual-element cylinder – it will be slightly more efficient than a single element). Combine the low price with the fact that there are no moving parts, and it's also the most reliable solar hot-water system you can buy.

There are some situations where diverted PV is either not a good choice or not possible.

You have a small roof. If you have small roof and large household energy consumption then you should *not* use that valuable roof space to add PV for resistive water heating.

You may have a large enough roof but want to reserve space for more solar PV panels in case you get batteries or an electric car in the future. In just a few years, I believe almost everyone will have batteries in their home and most people will be driving electric cars. These stationary and mobile batteries will require more solar PV panels to charge them affordably.

Diverted PV is not as space-efficient as solar thermal hot water (which has a much smaller total panel area) or solar heat pumps (which take up no roof space). If your roof space is limited, diverted PV is probably not for you.

Local network rules. Your local electricity network (your distributed network service provider, or DNSP) might have silly local rules that limit either the size of your solar PV panel array or the size of your inverter – or both. If there are local rules that limit the size of your solar system so much that you can't get all the solar PV you need, you should look at solar thermal hot water or a heat pump, as these are outside any network connection rules.

> **Online resource:** The rules do change frequently, so it's best to check. I've got an up-to-date list of each DNSP's rules here: solarquotes.com.au/sillyrules

What's your choice?

Tick which hot water option you're going for:

- ☐ Leave it alone – I'm happy with the condition, reliability, fuel, environmental impact and running costs of my hot water system. I'll just be getting quotes for solar electricity and won't touch the hot water.
- ☐ I'll get diverted PV including a new cylinder added to my solar system, and I want to rip out my gas hot water system (I plan to get off gas completely at some point).

STEP 3: Heating Your Water

- ☐ I'll get diverted PV with my solar system, and I can reuse my existing electric hot water cylinder.
- ☐ I'm going to get quotes for a heat pump and get the installer to configure it to run in the daytime – predominantly off solar PV.
- ☐ I don't mind splashing the cash, I don't want a heat-pump, and I don't want to use my PV for water heating. I'm getting old-school solar thermal hot water.

When you get quotes for solar, refer back to your choice so you can brief the installers about the hot water system you want.

Summary

- LPG-fuelled hot water is a very expensive way to heat water. Seriously consider replacing or boosting it with diverted PV solar hot water or a heat pump.
- If you currently have mains-gas-fuelled hot water, estimate how much you're paying for gas to heat your hot water. If reducing this cost by 70% to 80% is worth it to you, I recommend starting your transition to an all-electric home. Rip it out and replace with a heat pump or diverted PV.
- If you have conventional electric hot water, estimate how much you're paying for electricity to heat your hot water. If reducing this by 70% to 80% is worth it to you then get either diverted PV or a heat pump. If your electric cylinder can be re-used then diverted PV should be cheaper to buy, unless you are in Victoria where there are extra rebates for heat pumps.
- If you have an electric heat pump and you're happy with it, don't touch your hot water system but get at least 6.6 kW

of PV so the heat pump will have plenty of power to run in winter.
- If roof space is tight relative to the amount of solar PV you want to install now (or in the future) then go for a heat pump instead of diverted PV in the above scenarios.
- If you discover that your local electricity network won't let you connect a PV system as big as you need for your regular daytime loads, consider solar thermal hot water or a heat pump, as they use the least electricity.

STEP 4: Show Me The Money!

Now it's time to calculate the optimum size of solar system you need to maximise your returns. In theory, this could get very longwinded. Seriously, I could write *War and Peace* on this stuff. Hey, I'm an engineer.

In practice, because of the way the solar rebate works, inverter pricing and the overhead involved in 'rolling a truck' to do an install, the smallest size most installers will quote for is 3 kW.

And the largest system that most single phase households install – thanks to the local network rules – is 6.6 kW. If you want to go bigger, you'll need an 'export limiting' inverter and special permission from the network, or a three phase supply. If you have a three phase supply and a large enough roof, most people can easily get permission to go up to 20 kW of panels.

Sizing your system

At the time of writing, the typical price for a 3 kW, 12-panel system is about $4,500. That's $1,500 per kW for the first 3 kW of panels.

> **Online resource:** An up-to-date list of ballpark pricing is available here: solarquotes.com.au/cost

The typical price for a 6.6 kW, 24-panel system is $6,500. That means you're getting an extra 3.6 kW for $2,000. That's just $550 per kW for the last 3.6 kW of panels. This means that the last 3.6 kW of panels comes at close to a third of the price of the first 3 kW.

> **Sizing secret:** The last 3.6 kW of panels can be a third of the price of the first 3 kW.

In terms of sizing your solar system, this means that, as long as you are getting a reasonable feed-in tariff in your area, you have the space on your roof and you can find an extra $2,000, you're almost certainly better off getting a 6.6 kW system because it should provide a better return. I'll show you how to calculate the return so you can confirm this for yourself.

If you plan to heat your water with a heat pump or diverted PV then you should certainly not go smaller than 6.6 kW.

The only parts of Australia where the feed-in tariff is unreasonably low, in my opinion, are some parts of Western Australia, where you have no retailer competition. Here, the feed-in tariff, at the time of writing, is about 7c per kWh.

But I suggest that even you folks in Western Australia should go large because:

- an important benefit of a larger system is increasing self-consumption on cloudy days, and early and late in the day
- you'll thank me when you get batteries and an electric car in a few years, and
- Western Australia is one of the cheapest places in the country to get solar, so you'll probably pay less than your non-Westie countrymen and women for your solar system to start with

STEP 4: Show Me The Money!

Key point: If you have a big enough roof, my advice is to go for at least 6.6 kW of panels, no matter where you are in Australia and no matter how small your daytime consumption. It will almost certainly will give you a better return than a smaller system.

If you use more than 14 kWh in the daytime, you should go larger than 6.6 kW if your roof and your DNSP allow.

Am I allowed to add more than 6.6 kW of solar on a single phase supply?

If you have single phase supply, most DNSPs have an inverter size limit of 5 kW.

You're allowed to oversize your inverter by 33%, so you can have 6.6 kW of panels on that 5 kW inverter.

If you want more than 6.6 kW on a single phase, there is a way round it (DNSP permitting).

Ask your installer about using an 8 kW inverter, export limited to 5 kW. Because it's an 8 kW inverter, you're allowed up to 11 kW of panels. It will never send more than 5 kW to the grid, so it is often allowed under the 'maximum 5 kW' rules.

> **Online resource:** Many people worry about 'overloading their inverter' when the installer suggests a panel array that is larger than the inverter. This link explains why it is actually a really good design option: solarquotes.com.au/xl

What system size are you going for?

If your daytime usage is less than 14 kWh per day, and your roof is large enough, I suggest 6.6 kW. If your daytime consumption is more than 14 kWh, then, if your DNSP permits it, you should simply fill your roof with solar up to 10 or 11 kW.

My preferred solar system size will be _____ kW of panels.

Great – now we know what size system we want, it is time to look at the financials.

The different approaches to buying solar

Let's be honest: if you have a grid connection, you don't *need* solar. Having solar panels on your roof doesn't add any features to your home. The electricity from the panels doesn't run your appliances any better than grid electricity does.

Why do people buy solar? They do it for these three reasons:

1. Financial: to reduce their bills and save money.
2. Ethical: to take responsibility for the energy that their home uses, and to generate some or all of it virtually carbon-free.[7]
3. Material: Solar panels are cool, and you just want them!

If your motives are purely financial, it's important that you understand the economics of your investment in solar. This will allow you to make an economically rational decision about whether you should spend thousands of dollars on a money-saving device.

If your motives are purely ethical, good on ya! You think that if carbon-free generation can be added to a home for a reasonable

---

[7] About 30g of CO_2 per solar kWh.

STEP 4: Show Me The Money!

cost, you have a moral responsibility to install it and reduce the load on the predominantly coal-fired grid from your home. You're probably even pretty relaxed about exporting your 'spare' solar back to the grid for a low price, because your exported solar is reducing the impact of your neighbours' energy choices.

I fall into this category. I was building a home, and I wanted my house to be part of the solution. I wouldn't have felt comfortable in a home that didn't take advantage of the cheap, clean solar energy falling on its roof. I bought a good-quality system, with a back-of-the-envelope calculation that the payback would take from five to ten years, and I was happy with that.

But most Aussies are motivated by financial *and* ethical reasons. They intuitively know that installing solar is 'doing the right thing', but the financial reality is that it has to pay for itself relatively quickly, and they want to optimise that payback by buying the right-sized system for their circumstances. The rest of this chapter is going to walk you through the process of predicting the payback of an investment in solar.

Your payback is calculated from two things: what your system costs you, and the financial benefits it gives you.

The costs of buying solar

The dollar cost of buying solar is more than the sticker price on that shiny new system. Every owner will incur two types of cost:

1. The cost of financing your system.
2. The cost of maintaining your system.

The cost of financing your system

This cost depends on how you pay for your system.

Paying from your savings. Obviously, this incurs the cash price, but it also has an 'opportunity cost' that you should account for. This is the money you may be losing by not investing your cash elsewhere – for example, lost interest if it was previously in a term deposit.

Using finance to buy the system. Taking out a personal loan, or adding it to your mortgage, will incur interest costs and other fees.

A payment plan (frequently advertised as a no-interest deal). Although they use clever language and legal jargon to imply that the finance is low-cost or free, these deals are almost always the most expensive way to buy solar. They typically add thousands to the cost of the system to cover the cost of the finance, and the financier piles on various other fees.

Tip
Be wary of any deal that promises 'no interest'. In my experience, these 'deals' are best avoided.

The cost of maintaining your system

Annual maintenance. Good news – there is none! Panels more than 10 degrees from horizontal will clean themselves in the rain and there are no moving parts.

Five-yearly maintenance. Your system is in a harsh rooftop location and most systems carry high-voltage DC.[8] I recommend a proper

---

[8] Unless you have micro-inverters.

STEP 4: Show Me The Money!

electrical inspection every five years to make sure it is working safely and reliably. It is likely that your local electricity network will also insist that you do this every five years. Budget $200 for this.

Replacing the inverter. Even the best-quality inverter is unlikely to last as long as good-quality panels. Realistically, expect to replace your inverter every 12 years.

The financial returns from solar

You will reduce your bills in two ways, as we learned in Step 1 – through exports and through self-consumption. Let's have a quick refresher on the difference.

Exports. These are the earnings you get from exporting surplus energy to the grid. In general, you can earn between 7c and 16c for every kWh exported. These earnings are easy to see, as they are line items on your post-solar bills and are subtracted from your usage charges.

Self-consumption. This is solar energy that has been used directly by appliances in your home. You save whatever your tariff is: typically, 30c per kWh. I call these savings the 'hidden savings' of solar, because they don't appear on your post-solar bills. Other people call them 'behind the meter' savings, because the savings are invisible to your meter, or the grid – which is why you can't see them on your bill. Confused? Let's go through one of my bills to explain. I touched on this in Step 1, but it's an important concept so here are the details.

Figure 4.1 is a typical summer bill for my house (6 kW of solar, solar thermal hot water, powering five people and a small business):

101

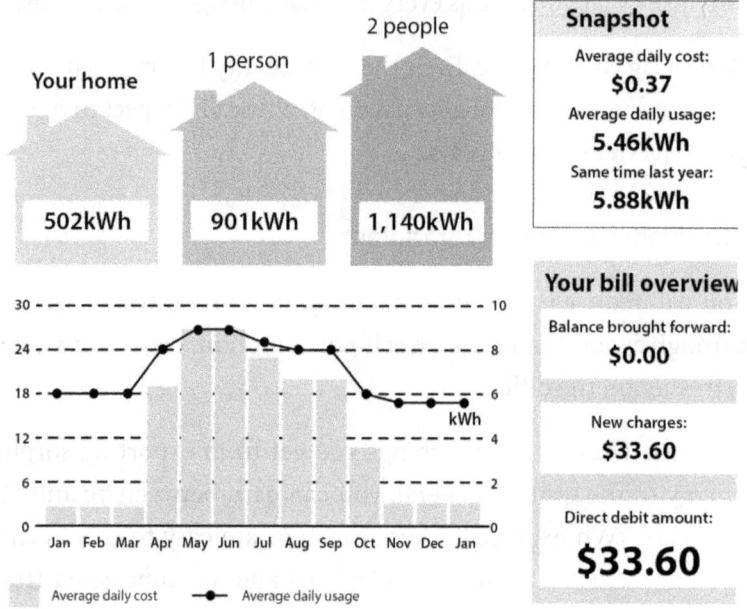

Figure 4.1 Summer electricity bill.

Obviously, at $33.60 for the quarter, I'm pretty happy – but how much am I actually saving compared with if I didn't have solar?

I don't have a pre-solar bill to compare, because I added the solar when I built the house. And even if I did have a pre-solar bill, the difference between the two would be a poor estimate of savings because usage and rates both change. In fact, it's common for people to use a lot more electricity after they install solar, because they worry less about their bills – so comparing 'before and after' bills can be misleading.

If we look at my bill, the savings from exports are there in black and white:

STEP 4: Show Me The Money!

New charges and credits

Usage and supply charges	Units	Price	Amount	
Non-Summer - 14 October 14 to 31 Dec 14 (79 days)				
Peak	260kWh	$0.289	$75.14	
Peak next	171kWh	$0.2946	$50.38	
Supply charge	79 days	$0.6379	$50.39	
Summer - 1 Jan 15 to 13 Jan 15 (13 days)				
Peak	43kWh	$0.3048	$13.11	
Peak next	28kWh	$0.3048	$9.31	
Supply charge	13 days	$0.6379	$8.29	
Other charges				
Payment processing fee			$0.95	
Payment processing fee			$1.67	
Total charges				+$209.24
Credits				
Solar Buyback*	2109kWh	$0.08	$168.72cr	
Solar Buyback*	348kWh	$0.08	$27.84cr	
Total credits				-$196.56cr
Total new charges and credits				= $12.68
Total GST				+$20.92
Direct Debit amount (includes GST)				=$33.60

Figure 4.2 Savings from solar exports shown on my bill.

I've been credited $196.56 for exporting 2,457 kWh (2,109 kWh + 348 kWh) of energy to the grid.

But how do I work out how much I've saved through self-consumed solar energy? Unfortunately, it's not that easy.

The first thing you need is a good energy-monitoring system. Later, I'll talk about energy monitoring in detail, and I'll suggest that you include it in your solar system. There are lots of reasons to do so – including tracking your savings.

The Good Solar Guide

Good monitoring systems show you how much energy your home uses and how much solar you generate, in five-minute intervals.

Figure 4.3 shows what happens on a typical summer's day in my house:

The spiky black and dark grey areas show my power usage over 24 hours – from midnight to midnight. The spikes just after midnight are the dishwasher heating the water and then drying the dishes at the end of the cycle. We put it on a timer because it's noisy and don't want it to disturb our evenings.[9] You can see the house start to wake up at 6am. Energy is used throughout the day. Mostly, it's the clothes washer, oven (cakes!), computers, and rainwater pumps. We put the kids to bed at 7:30pm and energy use falls off.

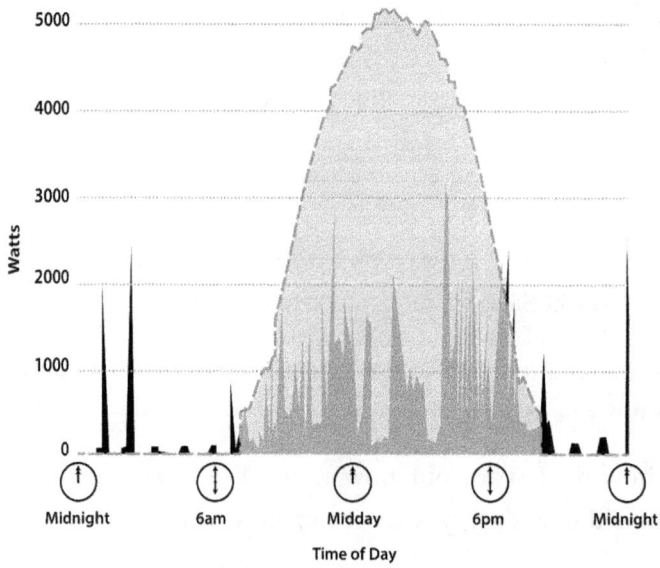

Figure 4.3 Breakdown of energy used in my house over 24 hours.

---

[9] We should rearrange when we pack and unpack the dishwasher so we can run it during the day, from the solar. See Step 6 – shifting loads.

STEP 4: Show Me The Money!

The dashed line shows solar generation. If the dashed line is above the spiky area then we're generating more than we're using. Any electricity being used at these times is self-consumed solar energy. I've shown this energy as dark-grey in Figure 4.3. Any excess solar gets exported. This is the light-grey area in the graph. The black areas show when I have to import from the grid.

The total amount of energy used by my house in the 24 hours shown is the black and dark grey areas combined. My monitoring system tells me this is 13 kWh of energy.

Of that 13 kWh, 10.33 kWh is in the dark grey area. This is my self-consumed solar energy. At $0.31 per kWh, that means I'm saving $3.20 per day through self-consumption.

My exports on this day add up to $2.51 (31.4 kWh x $0.08). That means most of my savings are coming from self-consumption.

My total savings for the day are $5.71.

If I extrapolate this to the 93 billing days covered by my bill and take into account that some days were overcast, the total savings come to about $500 for the billing period – $200 from exports and $300 from self-consumption.

If I extrapolate this to the full year (taking into account that winter generation is lower), then my savings for a whole year come to about $1,800.

Key point: Solar savings have two components – exports and self-consumption. We need to account for both in our predicted savings.

Calculating your predicted payback

First, let's collect the inputs to our calculation: We've already worked these numbers out and scrawled them on the worksheet:

System size you need: kW

Your daytime usage: kWh

Savings if getting solar hot water: kWh

Typical feed-in tariff available in your area: c per kWh

Typical usage tariff available in your area: c per kWh

Put the numbers above in the online calculator and you'll see your daily savings and first year's savings: solarquotes.com.au/simplecalc.

> **Online resource:** If you want to see how the calculator gets these answers, I'll walk you through the manual calculation here: solarquotes.com.au/maths

Getting more sophisticated

These simple metrics are good indicators of the value of solar for you. But they miss out a couple of important things.

Electricity price inflation

Most Australians recently suffered a large hike in electricity prices, and I believe further increases are likely.

Including rising electricity prices in your calculation will make

STEP 4: Show Me The Money!

the returns look even better, but this is a double-edged sword. Some salespeople can use big projected price rises to make the payback numbers look unrealistically good, so always ask what their assumptions are when they calculate your returns.

If you think electricity prices will rise then it makes sense to add them to your calculation. I've created a calculator that will do all the maths for you, and I'll show it to you in a minute.

First, let's talk about cash flow.

Cash flow

Looking at the simple payback, many people make a mistake. They think, 'Hmm, it will take four years to pay for itself. I can't afford to lose money for four years; cash is tight. I'll wait.'

This thinking ignores the cash flow of the savings. Let me explain.

Imagine you're looking at a $6,000 system that will save you $1,200 a year. That's a simple payback of five years. Another way to look at it is this: it will save you $100 a month.

If you can finance the system for less than $100 per month, then you're helping your cash-flow situation immediately. You'll have more cash in your pocket every month.

> **Online resource:** Learn the insider secrets of solar financing here: solarquotes.com.au/finance

Just be aware that, if you have tight cash flow every month (join the club!), the monthly savings of solar might be more important to you than the simple payback.

Comparing with savings or other investments

If you're in the lucky position of having cash to spend, but solar is competing with other investments (such as savings accounts or managed investments), you may want to compare the returns of solar with the alternatives.

In my opinion, the best way to do this is to calculate the internal rate of return or, even better, the modified internal rate of return. Confused?

The calculator I will show you will provide these metrics. First, I want to explain what they are, and their dangers and uses.

This is not an investment book, so I won't write a tome on investment metrics and describe how internal rate of return is calculated. If you're interested, this is a great primer: solarquotes.com.au/irr.

Suffice to say that the internal rate of return of an investment gives you an equivalent interest rate for your investment, but – and this is a biggie – it assumes that all your savings are reinvested at the same rate.

Let me explain. Suppose you put the numbers in my calculator. You're interested in your return over 25 years from a $6,000 solar system. The calculator pops out an answer of 20% over 25 years. That's the equivalent of saving a fixed $1,200 every year for those 25 years. (I'm ignoring electricity price inflation to keep things simple.)

Many people hear that they can get an internal rate of return of 20% on a $6,000 investment over 25 years and assume that it's the equivalent of plonking $6,000 in a term deposit with a 20% interest rate payable over 25 years.

STEP 4: Show Me The Money!

That's not true. Let's compare the returns of our imaginary solar system with those of the imaginary term deposit.

A term deposit of $6,000 earning annual compound interest at 20% for 25 years will provide a whopping balance of $572,377.

If you add up all the savings from the solar system ($1,200 x 25 years), and subtract the original $6,000 investment, you only get $24,000 in your bank after 25 years. That's quite a difference!

What gives?

Well, obviously, the second figure assumes you're getting zero return on your solar savings.

When comparing investments, we should assume that you invest the returns at a reasonable rate. No bank offers 20% interest.

The way internal rate of return works is that it assumes you invest the savings at the same rate as the internal rate of return. You'd have to find another 20% investment to plough your electricity savings into. Unless you own every house on the street, you're unlikely to be able to reinvest in solar, so you'll be keeping the savings in your bank account. (That's right, I'm assuming you don't spend the savings on chocolate.) A much more realistic rate is going to be 3% to 7%.

And this is where the concept of modified internal rate of return comes in. You can modify the internal rate of return by telling it what rate you believe you will be able to reinvest the money at.

My online calculator will find the modified internal rate of return for you. In our example, a $6,000 system returning $1,200 every year for 25 years gives a modified internal rate of return of 10% if we reinvest the savings at 5%.

In my opinion, that's the number you should use to compare with your compounding investments, such as term deposits.

In our example above, if you buy a $6,000 system, get a fixed saving of $1,200 per year for 25 years and reinvest all the savings in a 5% savings account, that's the equivalent of depositing the same $6,000 in a fixed-return 10% investment account for 25 years.

Here's the calculator: solarquotes.com.au/calc.

It will give you much more accurate answers, because it will take into account electricity price inflation, panel degradation and inverter replacement costs. It will also show you cash flow versus savings.

Put your numbers in there and see if the return from your chosen system size will give you a return that you're happy with.

Then, just for fun, if the system size you've chosen doesn't fill your roof (you can check here: solarquotes.com.au/roof) run the numbers for a system that will. If you get a decent feed-in tariff, I'm guessing that it makes sense to fill your roof with solar panels – if your DNSP and your funds allow it. Bigger is usually better.

Solar has a great payback for most homes, and now you know what it is for yours. You should now have a payback figure, cash-flow projections and a figure for lifetime savings if you decide to invest in solar.

Now I want to go back to batteries.

Do batteries make solar payback even better?

If you buy a battery then, instead of buying electricity from the grid, you use stored solar electricity whenever possible.

STEP 4: Show Me The Money!

Let's do some simple calculations – back-of-an-envelope stuff – to determine how much money you can save by doing this.

Remember calculating your night-time usage way back in Step 2? Now is the time to use that figure to see if batteries make financial sense for you.

If you read the article I linked to in Step 1 (solarquotes.com.au/batterymyths) you can probably guess what the results will be. But it's a simple calculation, so let's do it to see just how much money you can make or lose with a battery.

You save money with a battery by storing your excess solar during the day instead of exporting it to the grid. Then, as the sun goes down, your stored solar gets used to power your house instead of grid electricity. For each kWh of stored solar you use, you're saving what you would have spent on your usage tariff: around 30c per kWh.

What many people forget is that for every kWh of solar stored, you're forgoing your income from the feed-in tariff. Let's call that 10c per kWh. Also, a battery loses energy in inefficiencies every time it charges and discharges, due to a thing called 'round-trip efficiency'. Typically, you'll lose at least 10% of your energy.[10] That adds 10% to the 10c per kWh you lose from storing instead of exporting, so you're losing 11c per kWh.

For every kWh stored and used at night, the net amount you save is the usage tariff minus 110% of your feed-in tariff. In my example, that's 30c – 11c = 19c per kWh. To keep things simple, we'll ignore the battery inefficiencies in your calculation.

---

[10] I'm starting to see real world data that suggests many battery owners are actually losing 20-30% of their energy when they charge and discharge their batteries, so I'm being generous to batteries here.

Write your savings per kWh stored here or on the worksheet:

Net savings

= usage tariff (c per kWh) − feed-in tariff (c per kWh)

=   c per kWh

The best-case battery scenario is that you buy just the right size battery to get you through the night and then fully discharge it every night, drawing nothing from the grid. In practice, people typically use 80% of their battery capacity every day and reduce their grid imports by 70% to 90%. For the sake of this exercise, we'll assume the perfect conditions.

The ideal battery size to maximise self-consumption is the same as your night-time usage. So, for example, I use 3.6 kWh overnight so my ideal battery stores 3.6 kWh of energy.

Battery costs at the time of writing are close to $1,000 per kWh of usable capacity. They are going down by about 15% per year, and bigger batteries are slightly cheaper, but it's a good estimate.

That means my perfect battery will cost $3,600, fully installed.

Write your battery cost here or on the worksheet:

Your battery cost

= night-time usage (kWh/day) x $1,000

= $

STEP 4: Show Me The Money!

The nightly savings are the net savings multiplied by the nightly usage. In my case, that's 19c x 3.6 kWh = 68c per day, or $249.66 per year. This gives a simple payback (ignoring electricity price inflation) of 14.4 years.

The longest battery warranty I know of is 10 years, and I don't expect any battery to last much longer than its warranty. In my situation, that means I'm likely to have to replace the battery before it has paid for itself.

Do the calculation for yourself here or on the worksheet:

Daily savings

= net savings (c per kWh) x night-time usage (kWh)

=   c per day or $ per day

Yearly savings

= daily savings ($ per day) x 365

= $ per year

To find the simple payback of the battery, you divide the battery cost by the savings.

Write your battery payback here:

Payback

= Battery cost ($) ÷ yearly savings ($)

=   years

113

This is a simplified calculation. On the pessimistic side, it doesn't account for rising electricity prices. On the optimistic side, it doesn't account for battery degradation or charging/discharging inefficiencies.

> **Online resource:** My battery payback calculator takes all these things into account: solarquotes.com.au/batterycalc

Now we've looked at your battery payback from reducing your grid imports. Most people calculate a number between 14 and 25 years. Remember, the best battery warranties are for 10 years.

But there is another way to earn money from batteries right now that can improve the payback situation: grid support. Grid support is where you're paid to push power into the grid to help stabilise it.

It doesn't earn much, though. You can expect an extra saving of about $25 per year per kWh of storage. For example, if you're looking at a 4 kWh battery, you may be able to get another $100 per year of savings. Bear in mind, though, that it costs about $1000 to add the fancy control system to your battery system. Last time I checked, it also chews up gigabytes of data per month off your internet plan to send all your data to the company servers for number-crunching.

In my experience, grid support doesn't really move the needle in terms of making residential batteries stack up economically, but it's a feature that's fun to have, if, like me, you get your kicks from watching battery controllers perform their tricks. Also, they do – in a small way – help stabilise the grid, which should pave the way for more renewables to be integrated.

STEP 4: Show Me The Money!

There's a good possibility that batteries will be able to earn much more in the near future.

Will batteries become free to charge?

Thanks to wind power and low demand, we're heading towards low wholesale electricity prices late at night, and thanks to solar, we can expect low wholesale prices during the day. During the evening, when renewable generation is low and demand is high, we're looking at high wholesale prices with dispatchable gas, hydro and battery power meeting most of the demand.

As long as wind capacity expands, wholesale electricity prices may fall to zero late at night. Zero wholesale prices during the day will also be fairly common, thanks to solar. And wholesale prices during the evening will be as high as the owners of dispatchable generators can get away with charging.

Wouldn't it be nice if you could charge your batteries for zero cents per kWh and save yourself the peak evening rate?

I think this is likely to happen in the next five to ten years. Combine that with the steadily reducing cost of batteries, and batteries have a bright future. But right now, buying a decent-sized battery will add at least $10,000 to your outlay and make your payback worse, not better.

Summary

- If you can fit enough panels on your roof, get at least 6.6 kW of panels with a 5 kW inverter.
- If your daytime usage is over 14 kWh, fill your north-,

west- and east-facing roof spaces if possible. Most homes will accommodate about 10 kW – assuming your network approves it.
- You have worked out the returns of your chosen system size. You understand that if electricity prices continue to rise, the returns get will better and compound.
- If you are paying with savings, you've checked the returns compared to other investment options using my modified internal rate of return calculator.
- If you are financing the system, you've used the online calculator to check that installing the system will improve your cash flow.
- If you're going to buy a battery because you want backup or you love the thought of them, go for it. In general, though, batteries will not pay for themselves before their warranty expires, and they'll usually make the returns of the system as a whole worse. I suggest waiting for battery economics to improve before buying them.

STEP 5: Choosing Your Hardware

When you get quotes for solar, you'll be presented with lots of different choices of solar panels and inverters. How on earth does a non-solar expert decide which technology types to choose and which brands to go with?

This section will walk you through different technologies and brands, so you can make an informed decision about what panels will go on your roof for the next 30 years, and which inverter all your solar energy will go through.

First we need to understand how solar panels work together to power your home.

You've probably never sat down and thought about how electrical current flows through a bunch of solar panels, and you probably never thought you'd have to. But if you're going to make an informed decision about buying solar, you need to appreciate that a conventional solar panel system has one big shortcoming: if shade falls on just one panel, all the other panels in the same 'string' will be affected.

Let me illustrate the problem with a picture of some bird poo: see Figure 5.1. Here are three solar panels in a row, and one has been crapped on, reducing its power output by 50%. Notice how, although only one panel has a poo problem, all three panels have had their power reduced.

The Good Solar Guide

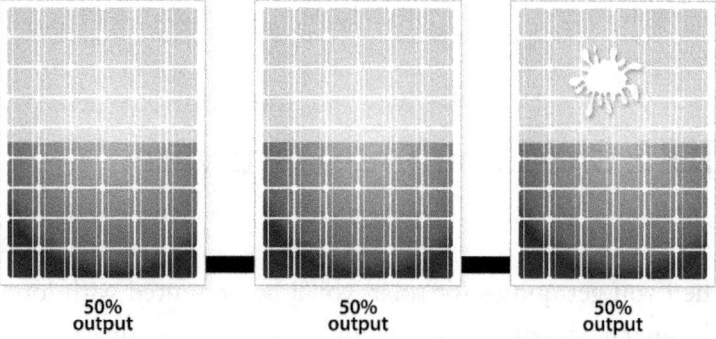

Figure 5.1 Three solar panels connected together; if one loses output, they all lose output.

This is what happens in a non-optimised solar system. You could have 12 panels in a single string, all being affected by a problem with a single panel. I'll explain exactly why this happens in a minute.

If you pay extra and opt for a 'panel-optimised' system (as in Figure 5.2), the problem is solved. That is, the bird-poo problem is confined to the pooey panel:

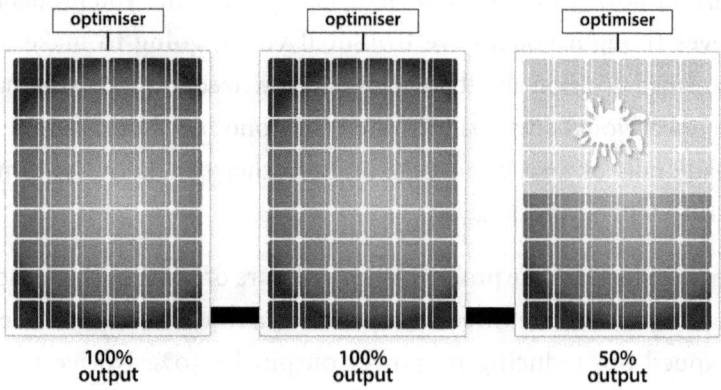

Figure 5.2 A panel-optimised system confines the 'poo problem' to just one panel.

STEP 5: Choosing Your Hardware

The first solar-hardware decision you have to make (because it can affect the type of inverter and panels you buy) is whether you want:

- a conventional system, known as 'string-level optimisation', where a problem with one panel in a string affects all the others, or
- a more expensive system, where a problem with one panel doesn't affect the others. This is called 'panel-level optimisation', which I'll shorten to PLO (with apologies to Yasser Arafat)

Online resource: To learn why solar panels bring each other down in a conventional system, head here: solarquotes.com.au/plo

Panel-level optimisation (PLO)

PLO uses smart electronics to optimise the power of each panel in a string independently. As explained already, if you have PLO and bird poo is affecting the panel, you only lose half of the pooey panel. The rest operate as normal.

PLO is useful for more than soiled panels.

- It reduces the effects of shade from trees and aerials.
- It optimises the power from panels if different panels are facing different directions.
- Even if your panels are all facing the same direction and you have no shade at all, PLO can increase energy yield by about 10%.

This last benefit is due to 'panel mismatch'. All panels operate at slightly different powers in the same sunlight due to manufacturing tolerances. They will also degrade at different rates. And between rain showers, some panels in a string will get dirtier than others. In a string system, this means that they'll all perform at the power level of the worst panel. PLO solves this.

You should get PLO if you have any shade on your roof, as the effects of shade on a conventional string are so drastic. If your roof space is so precious that you simply want to squeeze every last drop of power from your panels, PLO is for you.

If none of the above describes you then you'll be just fine if you get a regular string-level optimised solar system.

I've started the hardware section by talking about solar panel optimisation because PLO can affect the type of solar inverter or panels you buy.

If you want PLO, there are three options:

- Micro-inverters
- DC optimisers
- Maxim optimisers

> **Online resource:** You can see their pros and cons in detail here: solarquotes.com.au/compareplo

Micro-inverters and DC optimisers add about the same to the cost of a string optimised solution – expect to pay about $1,500 extra on a 3 kW install and $2,000 extra on a 5 kW install.

Maxim optimisers are the cheapest way to get PLO. Expect them to only add $750 to $1,000 to a 5 kW installation.

STEP 5: Choosing Your Hardware

Solar inverters

A solar inverter can be either a single central inverter (which is the size of a briefcase) or made up of multiple micro-inverters, the size of paperback books.

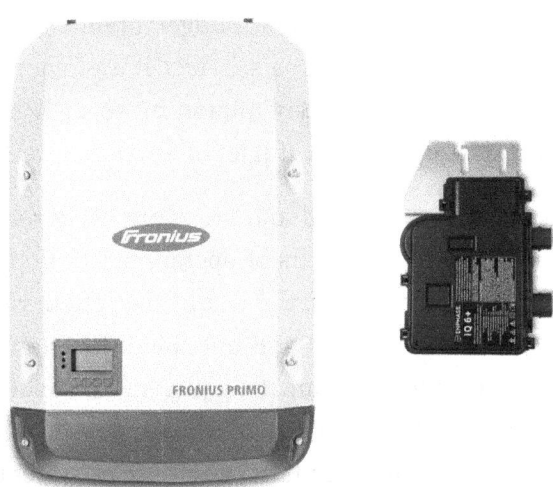

Figure 5.3 A central inverter (sometimes called a string inverter) and a micro-inverter.

The job of the inverter is to safely convert the DC electricity produced by the solar panels into 230 V AC electricity, which is what everything in your home uses.

This means that all your solar energy will go through your inverter or micro-inverters. This is important for two reasons:

1. **The efficiency of your inverter directly affects how much energy you'll get from your panels.** A 2% difference in efficiency in the inverter means a 2% difference in your energy production. That means the efficiency of your inverter is usually much more important than the efficiency of your panels.

121

In the next section, I talk about why, counterintuitively, less efficient solar panels give you the same amount of energy as more efficient solar panels. It sounds nuts, I know – but trust me on this.

2. **A central inverter works hard, gets hot and is the most common reason for solar systems failing in the first few years of operation.** Even the reputable budget brands have high mortality rates in the first three years. In fact, it was reported that Origin Energy recently took a $17 million hit to their business due to the cost of replacing an awful lot of Sharp inverters.[11]

Micro-inverters also work hard and are more likely than your panels to fail in the first ten years of operation. The good thing about micros is that if one fails, the rest of the system will keep working. If you have 20 panels, and one micro-inverter on one panel fails, the other 19 will hum along just fine, and you'll only lose 5% of your generation until the defective micro-inverter gets replaced. If a central inverter fails, your entire system stops generating.

When shopping for a central or micro-inverter, efficiency and reliability are your two priorities. I would say that reliability trumps efficiency. After all, if you have to replace the central inverter, your solar system will be offline for a few weeks, you'll lose an awful lot of energy and, if it's out of warranty, it will cost you at least $1,200 to replace. Replacing a micro-inverter, including labour, is likely to cost $300 to $400.

Reliability is directly related to quality, so let's start by looking at inverter quality.

---

[11] solarquotes.com.au/origin

STEP 5: Choosing Your Hardware

Inverter quality

You won't be surprised to learn that, in general, the more you pay for an inverter, the higher the quality and the more reliable it will be.

As an electrical engineer, I could spend the next 50 pages boring you senseless as I discuss how to design and build a reliable inverter that's suited to hot Aussie conditions. I'd be happy to do this, but I'm guessing you wouldn't be happy to read it.

All you need to know about inverter quality in Australia is shown in this chart:

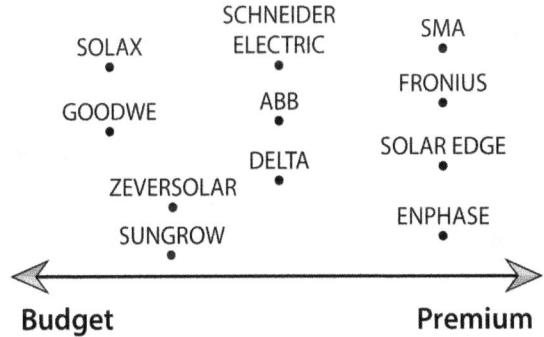

Figure 5.4 Reputable brands of inverter available in Australia.

Online resource: This chart will change as brands come and go – you can always find the latest version here: solarquotes.com.au/ichart

The inverter is the most likely component to fail in the first 10 to 15 years. This is because inverters work hard all day, and they do wear out.

That's why, if you're on a limited budget, I recommend prioritising a premium inverter over premium panels. The price difference

123

between opting for a premium inverter (such as Fronius or SMA) versus a budget inverter is $700 to $1,000 per system.

Enphase (the only micro-inverter brand worth considering, in my opinion) is the most expensive of the lot, because as we learned earlier, you need one micro-inverter per panel.

What should I buy?

One way of looking at it is this: a budget inverter is about half the price of a premium one. That means you can replace a cooked budget one that's out of warranty and be no worse off than if you'd bought a premium one. You will lose energy production while it's being replaced, though, and the installer's time will cost at least $200.

The other way to look at it is that a premium inverter is less likely to fail in the first 15 years. If it fails under warranty, the support and replacement should be fairly painless. Is that worth an extra $700 to $1,000? Only you can decide.

As you can probably tell, I'm a big proponent of buying quality and peace of mind if you can afford it. Like my granddad used to say: 'Buy it well, buy it once.' If the premium options are out of the question for you, I would strongly advise not going any cheaper than the brands on the left-hand side of the chart. There is some real junk out there.

Inverter warranties

The standard inverter warranty is 5 years. At the time of writing, SolarEdge offer a 12-year warranty and Fronius and Enphase offer a 10-year warranty; SMA will upgrade their warranty to 10 years for an extra $400.

STEP 5: Choosing Your Hardware

If I were buying a solar system right now – and I wanted a central inverter – I would buy SolarEdge or Fronius for the length of their warranty, or go with SMA with an upgraded 10-year warranty.

If I were to buy micro-inverters, I would go for Enphase – they come with a 10-year warranty.

Inverters for difficult roofs

You can skip this section if you have a simple roof!

If not, it's worth knowing that a little bit of magic goes on inside an inverter, which you probably aren't aware of. Despite what I said at the beginning of this section, the inverter has to do a lot more than convert the steady DC voltage into a jiggly AC voltage.

To get the most energy from a solar panel, you need to be careful about the electrical resistance that you connect to it. Too high and the current will stop; too low and the energy will be wasted.

It's kind of like your garden hose. If you want to make the water go as far as possible, you can put your thumb over the end. If you cover the hose too much, you just get a big mist – but there's a sweet spot, the 'maximum power point', where you leave a hole just the right size to get a powerful jet of water that shoots to the other side of the garden.

Inverters act like this. They vary their resistance until they get as much power as possible from all the panels they're connected to. This maximum power point changes depending on how much light the panels are absorbing at any moment in time. That means it is adjusted every second. The device inside the inverter that does this is called a Maximum Power Point Tracker (MPPT).

> **Online resource:** You can find the technical explanation here: solarquotes.com.au/mppt

I'm not just telling you this because I'm a geek and I think technology is amazing. It's because I want you to know that you need to be really careful if your panels aren't all pointing in the same direction. Panels that face different directions have differing amounts of sunlight hitting them so they have different maximum power points. That makes it impossible for a single MPPT to find the maximum of more than one set of panels at the same time if they are facing different directions.

If your home will need solar panels on multiple roof areas facing different directions, you might need a special type of inverter to ensure that you still get good system performance: a multi-string inverter (also called a multi-MPPT inverter), with one string or MPPT for each roof area.

Notice that I've used the word 'may', not 'must'.

This is because there is one scenario where you can get away with one MPPT on multiple roof areas – when your roof area faces east and west and has the same number of panels on each side.

Any other configuration is likely to need more than one MPPT if you don't want to cripple your system performance. A good installer will be able to look at your roof and work out how many MPPTs you need to maximise your energy yield.

Meanwhile, string inverters with more than two inputs are rare, so if you have a funky roof that points in more than two directions, you'll need more than one central inverter, or you should consider PLO.

STEP 5: Choosing Your Hardware

Inverters and panel capacity

A little known 'secret' of installing solar in Australia is this: whatever size of inverter (in kW) you get, you should (wherever possible) get 33% more panels connected to it. For example, if you get a 3 kW inverter installed, you should get 4 kW of panels on the roof. I've already touched on this when discussing system sizing.

Because this isn't common knowledge outside the solar industry, it is common in Australia for people to buy a solar system where the total capacity of panels in an array is the same as the capacity of the inverter. This has the advantage that you'll almost never lose energy because the panels never produce more power than the inverter can use, but this isn't much of an advantage.

Because panels rarely produce as much power as their rated capacity, it's possible to add extra panels with very little energy being lost. This extra panel capacity can help the inverter to run at a higher average efficiency, which can almost entirely make up for what does get lost.

When the total capacity of the solar panels is greater than that of the inverter, the panels are usually said to be 'oversized' or the inverter 'overclocked'. Because I think it makes a lot of sense, I tend to think of it as 'right-sized'.

You can overclock your inverter by up to 33% and still get financial help in the form of STCs as part of the solar rebate. (If you go even 1 W over the 133% limit, your application for STCs can be refused.)

The beauty of oversizing your panels with the current rebate scheme is that the rebate is based on your array size, not your inverter size. The extra rebate usually covers the cost of buying the extra panels. The only added cost to you is the labour to bolt the panels to your roof – typically, $400 to $500 for $1,000 to $2,000 dollars' worth of extra panels.

Oversizing your solar panels can be very cost-effective. Another real advantage lies in increasing your energy production when your local grid operator limits the inverter size you can install. For example, in many locations, people with single phase power are limited to installing inverters of 5 kW or less.

> **Online resource:** How to oversize your solar panels if you choose micro-inverters: solarquotes.com.au/oversizemicro

Batteries, inverters and surviving the zombie apocalypse

If you want to add a battery in the future, when batteries are cheaper, the good news is that this does not affect your inverter choice now. Any solar inverter can have batteries added later using a technique called AC coupling, which you can learn about here: solarquotes.com.au/accoupling.

If you want to get a battery right now, along with your solar, then you will be offered two options:

1. A single inverter that handles your solar and your batteries. This is called a 'hybrid inverter'.
2. A solar inverter for your panels and a separate battery inverter for your batteries.

I'm a fan of the second option, because it allows flexibility. If you want to upgrade your solar panel array, you only have to upgrade the solar inverter – and if you want to upgrade your battery, you only have to upgrade the battery inverter. When your solar inverter fails (it will, eventually) you only have to replace one, and you'll

STEP 5: Choosing Your Hardware

have more choice because it won't have to be compatible with your battery.

I've also found that the monitoring software that comes with many hybrid inverters is often pretty limited. I prefer the freedom to choose monitoring from a dedicated monitoring software company. We'll look at monitoring in more detail later in this chapter.

No matter what battery inverter set-up you're offered, you need to make sure it will give you the backup you expect. I touched on this when describing hybrid solar systems in Step 1. Now I'll dive into more detail, because it's important you get the backup you need when the grid goes down and the zombies come out.

Battery inverters and backup

If you're planning on buying batteries with your solar system because you want to be protected from blackouts, here's what you need to know to get a system that will meet your expectations.

Not all hybrid or battery inverters offer backup. To many people, this comes as a big surprise. Imagine you have a big, expensive battery on the wall and the grid goes down. Now imagine that the battery has been designed so that it cannot back up your home. You're in the same situation as the rest of the street – powerless. That would be embarrassing, especially if you've shown off your big, shiny battery to your neighbours.

Believe it or not, many battery systems sold today either have backup as an optional extra or don't offer backup at all. If you buy a battery without backup, you're buying it only for economic reasons – to store solar in the day and use it at night. The moral of the story is: you must tell any quoting solar companies that you want backup.

Most batteries will not back up your whole house – only certain circuits. Most reasonably-priced battery systems simply can't deliver enough power or energy to keep your home running as normal during a blackout. There are two reasons for this:

1. **Not enough power:** A typical battery system will have a 5 kW power limit. Your oven might be 3 kW and your ducted air conditioning might be 3 kW. Switch both of those on during a blackout and you've overloaded the battery inverter and tripped your power supply, and now you're fumbling around in the dark. If you haven't drained the battery of all its energy you should be able to switch the offending appliances off and restart the system, but worrying about tripping your battery inverter during a blackout isn't ideal.

2. **Not enough energy:** This is worse than tripping the power supply. Imagine you have a battery that can deliver 5 kW and store 10 kWh of energy. If (and that's a big 'if') it's fully charged when the blackout happens and your home is pulling 2.5 kW, your battery will be drained of energy in just 4 hours. But if you've only got LED lights, modem, fans, fridge, TV, radio and phone chargers on the backed-up circuits, these will probably pull less than 400 W. That gives you 25 hours of backup supply.

For these reasons, you'll have to decide which appliances will run off the battery and which you can live without. A good installer will decide with you which circuits you need – these are called 'critical loads', and only these will be powered by the battery when

STEP 5: Choosing Your Hardware

you're in off-grid mode. Usually, these include your lights, fridge, modems, office equipment and a TV. Let's be honest, how are you going to get through a blackout without *The Voice*?

Beware any solar company that claims their 5 kW battery will provide 'full backup' for your home. Unless you have a very efficient house with a low peak power draw, this is not going to happen.

Even if your battery or hybrid inverter offers backup, it might not be able to use solar to charge the batteries during a blackout. The nice thing about having a grid connection is that it acts as a buffer. If your batteries are full and your solar panels are generating more energy than your home can use, it's not a problem. The grid will soak up any excess solar and even pay you for it. Handy.

If the grid goes down, you lose that buffer. If your batteries are full and your solar panels are generating more than you can use, there's nowhere for the excess solar to go. If you can't somehow tell the solar panels to throttle back their power output, you'll get a big bang and your battery inverter may well start to send out flames.

For this reason, many battery inverters will cut power to the solar inverter during a blackout. This stops the solar panels doing any damage, but it means your batteries can't be charged. Once they're empty, there's no more power until the grid is up and running again.

Some battery inverters have been cleverly designed to communicate with the solar panels, so they can throttle them back when needed. If you want to be able to charge your batteries from the sun without a working grid connection, you need to get one of these.

Now, may I ask why you want backup?

1. **Because I like the idea of powering through blackouts**, and I'm confident that any blackout will last no longer than a day and a night. They don't happen often where I live, but the idea of blackout protection feels good.
2. **Because I have blackouts all the time and I'm sick of them!**
3. **Because, although blackouts are infrequent, they seem to be getting worse.** I'm worried about much longer outages. I don't trust the government to get supply back the same day – and I want to be able to keep the lights on through outages that last for days, charging my batteries by day and draining them overnight. If the zombie apocalypse comes, I want to be prepared.

If you answered 1, you probably don't need to charge your batteries without the grid.

If you answered 2 or 3, you need to ask for a system that will charge the batteries without the grid. Make sure any quotes specify this as a feature in writing.

Solar panels

Have you ever stopped to consider what solar panels actually do? Okay, they turn sunlight into electricity – most people appreciate that. But let's stop for a second and quantify what's really going on.

A typical 1 x 1.5 m solar panel (the size of a large flat-screen TV) will produce 250 W of power in strong sunlight. In a typical day, this equates to about 1 kWh of electrical energy generated.

STEP 5: Choosing Your Hardware

Each solar panel on an Aussie roof will provide an average of 1 kWh of energy per day. How can you visualise what that means?

Consider a middle-aged man in Lycra riding his bike. He's putting in a real effort. His heart is pumping, he's red-faced and sweaty, and you can smell him from the next street. If he's in very good shape, he may be pumping out an average of 200 W through his legs to propel him along. If he keeps this up for five hours, he'll have expended: 5 h x 200 W = 1,000 Wh. That's 1 kWh of energy.

That's right: 1 kWh of energy is the same as one very fit person cycling really freaking hard for five hours non-stop.

Jump off the sofa and have a look at one of the solar systems on your street (17% of homes have solar in Australia, so I'm assuming there are some near you).

Picture the fact that, today, each solar panel on that roof will produce the same amount of energy as five hours of hard slog on a pushbike. If the homeowner bought quality, they'll do that every day for the next 30 years, using no fuel and needing no maintenance. If you were to do the job of just one of those $250 panels yourself, it would be a full-time job – and imagine the food bill!

Solar panels are now so commonplace that we take them for granted. We shouldn't. They are truly amazing, and I strongly believe that we have a duty to only buy high-quality ones that will last for at least 30 years. It breaks my heart to see cheap, low-quality panels sent to landfill after just a few years, when for a few hundred dollars extra the homeowner could have bought panels that would have lasted for three decades.

The importance of solar panel quality

Thirty years on an Australian roof is a big ask. Heat, humidity, hail, possums, monsoonal rain and even cyclones – they all do their best to destroy those panels as they go about their business generating electricity, day in, day out, saving you tens of thousands of dollars over their lifetime.

To last 30 years, a solar panel has to be incredibly well made with excellent-quality materials and components. The cheapest solar panels, made to a price, will be unlikely to last that long.

Solar panel construction – at a high level – is simple. You sandwich 60 or more solar cells between two clear polymer sheets. Then place glass on top, a thicker, opaque plastic sheet underneath, and frame the lot with aluminium, as in Figure 5.5.

What could possibly go wrong? Lots.

I'm not going to bore you with an in-depth study of all the ways a solar panel can fail: if you buy well, you're highly unlikely to see any of them.

But just appreciate that the guts of a typical solar panel are made up of 60 beermat-sized, brittle cells less than 1 mm thick with many amps of current going through them for more than eight hours a day. They sit on your roof exposed to the Aussie sun, storms and heatwaves.

STEP 5: Choosing Your Hardware

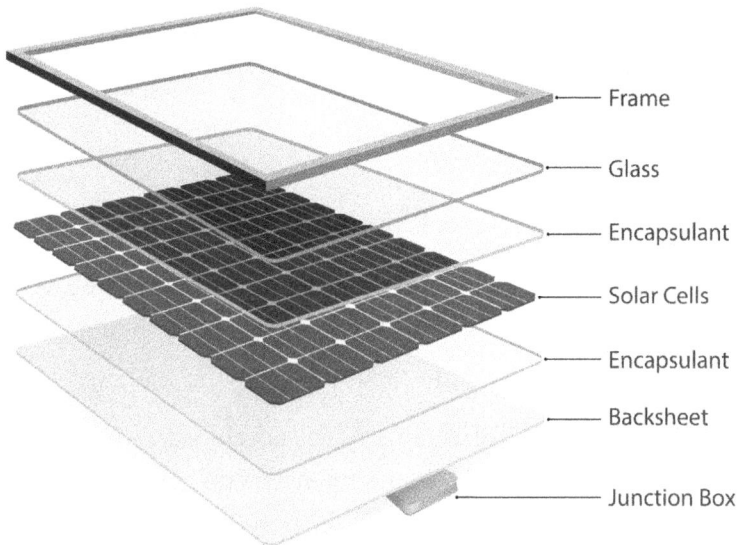

Figure 5.5 Composition of a conventional solar panel.

It's easy to build a panel that works well and looks good when it comes out of the box. The proof comes after it has been on your roof for two or three years. If any of the materials are substandard or they haven't been tested for Aussie conditions, the panel slowly starts to self-destruct – usually due to temperature stress, UV degradation, water getting in, or all three.

If this happens in a solar panel with up to 600 V of DC going across it, the results can quickly get nasty.

Unfortunately, there are dozens of solar panels offered by brands in Australia that appear to be made for a quick buck. They are made to a price, and that means cutting corners with materials. They're highly unlikely to last five years – never mind thirty.

And to compound the problem, most buyers wouldn't know a good solar panel brand from a crappy fly-by-night one – and why should you?

Luckily for you, there is a solution.

The almost foolproof way to avoid a crap solar panel

Understanding the quality and reliability of a solar panel and the company behind it isn't easy. The brands are mostly unfamiliar, and all solar panels look practically the same.

For this reason, the solar industry has developed a 'tier ranking' system to help guide buyers and financiers on which solar panel manufacturers are 'bankable'.

Tier 1 solar manufacturers are the most bankable, and Tier 3 the least. 'Bankable' means that a solar-farm developer using that manufacturer's panels is likely to be offered financing by banks for the millions of dollars they need to build the farm.

Banks don't usually lend millions of dollars without doing their research, so the idea is that if you choose a Tier 1 panel manufacturer you're buying a brand that the banks trust enough to lend millions of dollars against.

For obvious reasons, a panel manufacturer's ranking is a matter of great pride and is highly valued. It's a complex task to rank the many hundreds of manufacturers and a lot is at stake. Meaningful rankings are published by a small number of independent industry analysts. Due to the effort involved and the importance of the information, the rankings are generally not openly published. Instead, they are sold as industry intelligence. Last time I checked, it cost $30,000 per year to subscribe to the most popular list: the

STEP 5: Choosing Your Hardware

Bloomberg Tier 1 Rankings. Ouch. To make it harder for us Aussies, these expensive reports are based on an international ranking, not an Australian ranking.

To add another level of complexity, a few non-Tier 1 manufacturers sell what I believe are top-quality panels in Australia. Why aren't they Tier 1? Generally, it's because they're 'boutique' manufacturers who contract panels for the Australian market using Tier 1 manufacturers. In one case (Tindo Solar), the company is an Australian-based manufacturer who is simply not big enough to be classed as Tier 1, despite very high-quality panels.

All this means that the international Tier 1 lists – although a good tool – are not perfect for Australian purposes. They miss some great brands off, and they include some big brands that I wouldn't recommend due to poor Australian support or quality that isn't suited to Aussie conditions.

You should also be aware that, as I've described, Tier 1 rankings are not based on panel quality directly. The ranking organisations don't even look at the solar panels or their manufacturing lines to gauge the quality. I do consider them as a proxy for quality, though: if a panel manufacturer is always being used on massive solar farms by highly experienced developers, they're highly likely to be good quality.

You can't trust the Tier 1 lists blindly. They need to be taken with a pinch of salt and an eye to local Australian circumstances. For that reason, I've put together a chart showing Tier 1 and exceptional Tier 2 panels that I consider worth buying. The simplest way to avoid putting crappy panels on your roof is to make sure you use a panel on my list – see Figure 5.6. If your installer can't provide one of these brands, you're venturing into the unknown.

Panels to the left are cheaper, but they're likely to produce 5% or so less energy and not last as long as the panels on the right. In general, the panels on the right are also more efficient, which means they'll take up less space on your roof for every watt generated.

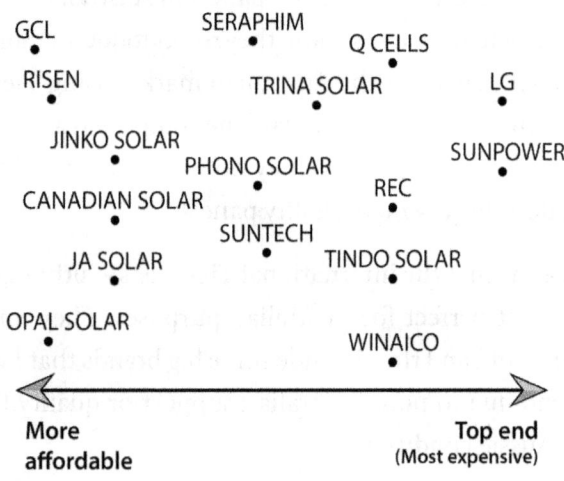

Figure 5.6 Reputable and well-supported brands of panel in Australia.

Online resource: Find the most up-to-date chart here: solarquotes.com.au/pchart

If you have been offered a panel **not** on this list, do the following:

1. Go elsewhere – there are hundreds of good solar installers in Australia who do offer these brands!
2. If you really want to use the panel brand you've been offered, shoot me an email and I'll give you my honest opinion: support@solarquotes.com.au. My response will most likely be: 'I've not heard of this panel brand – so I can't tell you if it's any good or not.'

STEP 5: Choosing Your Hardware

3. Don't take the solar company's word for it – 'This is a Tier 1 panel, mate, don't you worry!' It's common for the ultra-cheap mobs selling Tier 3 and 4 junk to claim their panels are Tier 1 and better than all the rest to get the sale. If it isn't on the chart, be extremely careful. Think about why the salesperson might be pushing an obscure panel brand. It's because they've got a good deal on it. The only reason to take a risk with obscure panel brands is price, so the salesperson's judgement is anything but objective.

When it comes to the chart, there's no 'master list' of reputable panels that everyone agrees on. This is my list based on extensive research and, yes, my personal opinion. There will be good panels that are not on this chart, that I simply am not familiar enough with to offer an opinion.

There is no guarantee that a panel from a company on this list won't fail during the warranty period. Even the best companies make mistakes, and today's good solar panel companies have had quality problems in the past. If you do need to claim on warranty, though, these companies are more likely to have the infrastructure in place in Australia to handle your claim easily.

Neither is there any guarantee that these companies won't go broke. Big companies do go broke, for lots of reasons, and companies ranked as Tier 1 have gone bankrupt before.

What I'm saying is that the chart is an indication of good quality and reasonable financials. It's the best way I know of choosing a good solar panel and reducing the risk of buying a low-quality panel from a company that won't be around to service the warranty. Buying a panel brand that isn't on the chart is taking a big risk to save a few hundred dollars on a system that will be on your roof for decades.

Now I've (hopefully) convinced you to only consider well-known brands, the big question is: 'How much extra are these panels going to cost me compared to the cheap ones?'

Here's the thing that blows me away. On a good-sized 5 kW system, the difference in price between the cheap panel of unknown quality that could have been put together by anyone and the Tier 1 panel built by robots in an internationally accredited, quality-assured, spotless factory (drum roll, please...) starts at about $600 on a $6,000 system.

That's an extra $600 for the peace of mind that your panels will generate safe, clean power for the 30 years they're on your roof. You may move home a couple of times in that time, but I strongly believe we all have a duty of care to every family that may own a solar-powered home.

It blows me away that, for the sake of $600, people are choosing Tier 2, 3 or 4 panels and risking their investment going up in smoke (sometimes, literally).

Every solar installer I know sleeps much better at night when they install quality panels than when they fit the budget panels that they may carry as an option for the people who insist on screwing them down to the last $100.

If that's true, why do some solar companies only offer the Tier 3 (and worse) junk? Because, unfortunately, they're in a race to the bottom. They need to have jaw-droppingly low prices on their TV and newspaper ads to get calls from people who are looking for the lowest prices, so they'll cut costs wherever they can.

The problem for these companies is that in their race to the bottom, they might just win.

STEP 5: Choosing Your Hardware

Detailed solar panel specifications

Solar panels have many specs you can pore over – efficiency, temperature coefficient, performance ratio, and multi- or monocrystalline. Really, none of these things matter if you're getting a good brand with a strong Australian presence to handle any warranty claims. The performance of the good brands is so close that you won't notice the difference in day-to-day operation. What you will notice is if the panels fail and you struggle to get a replacement under warranty.

So you now have a list of brands, so how do you choose between them?

If you find a good installer (I'll show you how, later), they'll have brand preferences. They'll usually install two or three brands: budget, mid-range and premium choices. They will have chosen their brands based on personal preference, experience, local climate, and, yes, what deals they can get with the panel manufacturer or importer.

If they've been around for a while and they plan to be around for a while longer, they want to sleep at night for the duration of the 25-year warranty. That means they're unlikely to offer a panel that they don't believe in.

If you find a genuine, local installer with plenty of good reviews online, offering reasonable (not jaw-droppingly low) pricing and panels that are on the chart, I would trust their judgement.

I could write thousands of words about solar panel specs, but in this section I'm just going to cover two key considerations.

Spec 1: Efficiency (and the only time it matters)

Salespeople love to wax lyrical about solar panel efficiency, but the truth is, it's not that important to most buyers.

It makes sense that the most efficient panel must be the best, giving you the most power and, therefore, the greatest return on investment, right?

If you're buying a washing machine or car then this has some truth. The more efficient your washing machine or car, the cheaper they'll be to run.

Does that mean that the more efficient your solar panel, the more energy you'll get?

No.

The efficiency of a solar panel measures how much of the sunlight that falls on it is turned into electricity. It varies from 15% to 23%, depending on the model and manufacturer. If solar systems were sold by the physical size of the panel array, the system with the highest efficiency would give you the most energy. But solar systems are not sold by their physical size; they are sold by their peak power rating in kW.

This means a 6 kW solar system with 15% efficient panels will require 24 of them, while with 20% efficient panels only 18 will be needed. Both systems will give you the same amount of energy, but the lower-efficiency system will take up more space on your roof because you'll need more panels.

Higher-efficiency panels are often more expensive per watt than lower-efficiency ones because they use the latest technology, and that costs money to develop.

STEP 5: Choosing Your Hardware

That means the only economical reason to pay more for a higher-efficiency system is if your roof isn't big enough for the system size you need.

As I mentioned before, there's an argument to be made (and the super-efficient panel manufacturers, LG and Sunpower, love to shout it from the rooftops) that in the near future, roof space is going to become much more valuable. As batteries become cheaper, people will install home batteries and electric cars, all of which need charging. With limited roof space, the best way to prepare for this is to buy the most efficient panels you can now, leaving space for even more as you need them.

I think that's a valid argument for people with an eye to the future – perhaps five to ten years away.

Let's run the numbers:

- **Home batteries:** A typical grid-connected Aussie home needs a 10–15 kWh battery to meet peak demand and get through the night on solar. It should have at least 6 kW of solar on the roof to be able to charge the battery through the day while still powering the home – even in winter. If you can't fit 6 kW of 'standard efficiency' (15%–17%) panels on your roof and you expect to buy a home battery at some point, spend more on super-efficient panels.
- **Electric cars:** At the time of writing, the most affordable electric car with a reasonable range – the Chevy Bolt (not available in Australia, unfortunately) – has a 60 kWh battery. The average car in Australia goes about 40 km a day, which would drain the car battery by about 7 kWh. To reliably generate an extra 7 kWh from solar, even in winter, you'll want another 3 kW of panels.

If you believe you'll still be in the same house and that you'll have a home battery and an electric car in the near future, use panels with a high enough efficiency to fit 9 kW of solar on your roof. You may only be buying 3–6 kW at the moment, but I believe that electric cars and widely adopted home batteries are less than five years away, so my advice is to use panels that will allow you to upgrade to at least 9 kW in future.

The higher efficiency, more expensive panels are those towards the right-hand side of the chart.[12]

Spec 2: Temperature coefficient

Australia gets hot. Contrary to popular belief, a solar panel's efficiency reduces as it heats up. Solar PV panels love light but hate heat. When the sun is beating down and the panels get hot, their power output drops by around 0.4% for every degree the panel gets above 25°C.

Note that this is the temperature of the actual solar panel, which is usually about 25 degrees above the air temperature. On a 40-degree day, when the panel is 65°C, the power output will be reduced by about: 40°C x 0.4% = 16%.

Some panels perform better in the heat than others. If you have to choose between a couple of panel models and you want to see which ones do best in the heat, you can compare a number on the specification called the 'temperature coefficient of Pmax'. Be careful to look at that number, not any of the other temperature coefficients.

---

[12] Be aware that there is one exception to the 'more expensive, more efficient' rule: Tindo panels. Tindo are Australian-made panels that are expensive but low-efficiency. This is not a criticism: they have been designed to prioritise durability over efficiency in Australian conditions.

STEP 5: Choosing Your Hardware

Temperature Characteristics	
Nominal Operating Cell Temperature (**NOCT**)	45±2°C
Temperature Coefficient of Pmax	-0.45 %/°C
Temperature Coefficient of Voc	-0.34 %/°C
Temperature Coefficient of Isc	0.050 %/°C

Figure 5.7 Excerpt from a solar panel data sheet showing three temperature coefficients.

The smaller this absolute[13] number, the better the performance. For example, a temperature coefficient of -0.42% per °C is better than a temperature coefficient of -0.49% per °C because you'll lose less efficiency for each degree the temperature rises.

Online resource: There is a detailed discussion of this on my blog: solarquotes.com.au/temp

As with most specs, good solar panels all have similar temperature coefficients, so there's no need to go crazy about hunting down the lowest. A really good temperature coefficient may give you a few per cent more energy over the year – but are you really going to notice?

To summarise, my advice is simple: buy a panel on the chart, preferably from a reputable local installer. If you can afford to buy something towards the middle or right-hand side of the chart, you'll be rewarded with around 5% more energy from your roof and in my opinion, less likelihood of making a warranty claim in the decades ahead.

---

[13] In maths, 'absolute' means 'without the minus sign'.

> **Tip**
> If you need a solar panel that will cope with special environments, such as salt mist (if you live near the sea), ammonia (near farm animals) or hail, this blog post explains what certifications to look out for and how to ensure your panels really are certified: solarquotes.com.au/certifications.

A monitoring system for your solar

The final piece of hardware I urge you to think about is a monitoring system.

A monitoring system will measure your:

- energy consumption
- solar production
- grid exports
- grid imports, and
- battery charge and discharge power

In general, the monitoring system will send this data to the internet every few seconds. You'll be given a login where you can see various graphs and numbers that tell you what's going on.

For an engineer like me, poring over graphs and numbers is heaven. But I totally get that, for most normal people, it sounds as much fun as doing your taxes. So why am I recommending that you invest in a monitoring system?

STEP 5: Choosing Your Hardware

Reason 1: Timely alerts when your solar system trips off

Even if you never look at your dashboard, a good monitoring system will alert you as soon as anything goes wrong with your solar system.

A typical Aussie home will save at least $500 per quarter with solar, often much more. Most homes get their energy bills every three months. If your solar inverter has an error and shuts down, you're not likely to notice for up to three months, when you open your bill and fall off your chair. You see, a solar system just sits there and works. If it stops working, the grid simply steps in and provides the power instead of your panels. You'll never know until the bill arrives.

From the emails I get, I can tell you this happens all the time. Imagine getting used to $50 bills and then getting hit with a bill for $550 or even $1,050. It makes you feel sick to the stomach when it happens.

With a good solar monitoring system, you'll get an email or SMS within a couple of days of the system going offline. Then you can call your installer and get the system back online ASAP.

Reason 2: Timely alerts if your panels' performance drops

Solar panels come with a minimum performance guarantee of 25 years, but how do you know if their performance ever drops below that guaranteed minimum? A good monitoring system can check that for you. It will know the correct performance associated with the direction and angle of your roof and even the local weather, and it will alert you to poor performance. This is much more sophisticated than telling you if the system has tripped off; the good systems can advise to within a few per cent whether your system is performing as it should be.

You're paying thousands of dollars for those panels to be installed, so it makes sense to be alerted if you stop getting the performance you paid for.

Reason 3: Timely alerts if your inverter's performance drops

We learned that the efficiency of your solar inverter is directly proportional to your power output. If its performance drops then your savings fall too. A good monitoring system will alert you as soon as there is a problem with your inverter so you can get it repaired or replaced.

Some inverters come with built-in solar monitoring. I recommend a third-party monitor to keep the inverter manufacturers honest. Using the inverter manufacturer's monitoring to advise you if you should make a claim on their warranty is putting the fox in charge of the hen house.

Reason 4: Safety

As our homes turn into distributed power stations, the consequences of electrical faults can be more severe. As a former electrical engineer, I can tell you that big power stations have all sorts of monitoring systems to alert the operators to faults that may be early indicators of dangerous or costly equipment failure. I believe that your home should have similar protection.

A good monitoring system can detect all sorts of fault conditions using funky algorithms. These conditions can be warnings that something is close to failure. For example, a detected fault in a

STEP 5: Choosing Your Hardware

solar panel may be a precursor to panel breakdown which, if left, could lead to the high DC voltage arcing and causing a roof fire.

A good monitoring system makes your solar safer.

Reason 5: Understanding exactly why your bill has gone up or down

I get emails all the time from solar owners whose latest bill is unexpectedly high. The first thing we check is that the solar system is working. Often, it is. The only rational explanation for the high bill could then be:

- the performance of the solar reducing
- consuming more energy
- changed tariffs, or
- a billing error

Working out the cause requires maths, accounting ability, engineering skills and lots of time. Most people give up in frustration.

A good solar monitoring system can give you the reason in two minutes, so you can take the right action to get your bills back down.

Reason 6: Optimising your retail offer

Your area probably has at least a couple of dozen electricity retail tariffs to choose from. With the various discounts and add-on charges, the tariffs are confusing and difficult to compare. A good solar monitoring system can continuously track your usage, compare it with current offers and alert you when a retailer offers a new tariff that will save you money. Take that, electricity retailers! This alone should pay for your monitoring system many times over.

Reason 7: Energy efficiency

If you're interested in improving your home's energy efficiency, the insights a monitoring system can give you will really help. Here's an example: your 'always on' power is the minimum power your house draws from the grid – probably in the middle of the night. It is the level of power consumption that you never drop below. This number reveals how big your 'standby loads' are, so you can identify appliances that suck power even when you're not using them.

Monitoring can also help you discover which appliances you could start using in the daytime so they use solar instead of grid power – increasing your self-consumption ratio.

Which monitor should I buy?

As mentioned already, many inverters come with monitoring. Almost all of them will let you monitor their solar production remotely these days, and some include consumption monitoring. I'll let you in on a little secret: most of them suck compared to a good third-party monitoring system. Why? Because they're made by inverter manufacturers, and the software is a side gig.

Some of the inverter manufacturers that have reasonable monitoring software include Fronius, SMA and Enphase. Even so, from what I've seen, none of them can alert you to a failure much more subtle than total system shut-down, except Enphase, whose system should tell you if a single micro-inverter fails.

A solar monitor from a solar monitoring company, which does nothing but think about how to make their monitoring software

STEP 5: Choosing Your Hardware

better every day, is the best way to keep tabs on your super-duper solar (and perhaps battery) system. A third-party system is the only way to get all the features I described above.

The only third-party system I recommend at time of writing is an Australian system called Solar Analytics. It does everything mentioned above except 'Reason 6: Optimising your retail offer', but they assure me that feature is being worked on.

The downside of this package is that, to unlock its full potential, you have to pay $6 a month. The upside is that, if you get it installed with your solar system, you should only be looking at about $300 for the hardware.

If you can wear the monthly fee, I highly recommend adding Solar Analytics to any solar system you buy. To put the $6 a month in perspective, if it alerts you to a warranty issue on your inverter or panel array before the warranty expires, it could pay for itself for twenty years.

I think of a monitoring system as cheap insurance on a critical service, and consider it an essential addition for any savvy solar buyer.

Summary

Solar panels

- Solar panels should last for 25 to 30 years. Australian roofs are harsh environments, so quality is key. The difference in price between 5 kW of reputable panels and 5 kW of junk can be as low as $600, so buying quality doesn't have to break the bank.

- If you choose one of the brands on the chart in this section (or better – the up-to-date one online: solarquotes.com.au/pchart), you'll get what I consider to be a reputable brand. Although there are no guarantees in life, I believe these brands are much more likely to last and that it is more likely the companies will be around to service any warranty issues.
- Unless my online tool shows that your roof is too small to fit enough 'regular efficiency' panels for your system, you don't need to buy super-efficient, super-expensive panels (unless you simply like owning the best).
- As for the other panel specifications, don't sweat it. Good panels are all similar in this regard.
- Make sure you understand solar panel warranties, which I go through here: solarquotes.com.au/warranties.
- If you buy solar panels without PLO then if one panel in a string has reduced output, it can drag all the panels down to its level. PLO is essential if you have partial shade, or a complicated roof that faces in more than two directions.
- PLO can be achieved with micro-inverters or DC optimisers, but it will add $1,500 to $2,000 to the cost of a 5 kW system. PLO can also be achieved with solar panels that have Maxim optimiser chips embedded. This only adds $750 to $1,000 to the cost of a 5 kW system.
- Make sure your panels have all the certifications you need for special conditions, such as salt-mist resistance.

STEP 5: Choosing Your Hardware

Inverters

- The inverter is the most likely component to fail in the first ten years, so focus on quality and go for brands on the right-hand side of the chart in this chapter (or better – the up-to-date one online: solarquotes.com.au/ichart). The difference between the budget and premium brands is $700 to $1,000 for a 5 kW inverter.
- Get a 10-year warranty with your inverter – even if it costs you an extra $400 to extend it from 5 years.
- Get 33% more panels than your inverter is rated at. This is the most cost-effective way to buy solar and it won't negatively affect performance in any meaningful way.

Batteries

- If you plan to add batteries later, there's no need to buy a special inverter. You can easily add batteries to *any* solar inverter with a technique called AC coupling.
- If you plan to get batteries and an electric car in the future, look for panels with an efficiency that will allow you to fit at least 9 kW on your roof, even if you're not installing 9 kW now. Don't be afraid to use all roof directions.
- If you want apocalypse-proof backup, make sure any quoted system states that the batteries can charge from the solar during a grid outage.

153

Monitoring

- A good third-party monitoring system will alert you almost immediately if your system fails or underperforms. Don't wait for your quarterly bill to learn that your solar savings have stopped or reduced.
- Third-party monitoring will monitor your consumption so that you can understand what appliances are still using grid electricity and identify opportunities to reduce grid consumption.
- A good monitoring system will tell you what your optimum battery size is and how much it will save you, so you can invest in the right battery at the right time.
- In the near future I expect third-party monitoring to advise you proactively if you are on the best tariff.
- I personally consider good third-party monitoring an essential addition to any solar system.

STEP 6: Getting Quotes

The fact that you have started to read a chapter called 'Getting Quotes' tells me that you've decided solar is for you. Now we get to the pointy end. It's time to go out and get quotes. To prepare you for that, this step will arm you with the following knowledge:

- The five types of solar installer: who to avoid, and who will be most likely to deliver a system that makes you happy.
- Three different ways to find three reputable installers to give quotes.
- Whether you should invite the quoting companies to your home before getting the quotes.
- What a good-quality, comprehensive quote should include.
- Terms and conditions to look out for.

Making sure you understand these five things is the best way I know of ensuring you get a well-installed, reliable system that gives you the savings you expect.

Types of solar installers

In my experience, there are five types of solar installation company in Australia:

- Type 1: Crap solar
- Type 2: Cheap but decent solar
- Type 3: Happy medium
- Type 4: Expensive (craftsperson)
- Type 5: Expensive (rip-off)

Obviously, I recommend avoiding Type 1, crap solar, and Type 5, rip-off solar. Types 2, 3 and 4 are the ones I try really hard to refer through my website, SolarQuotes®. Each of those three types suit different buyers.

If you want a budget system with no frills, you live in a metro area and you have a straightforward installation, Type 2 could be the best match.

If you want a good mix of quality and price, a non-standard install and some extra functionality, Type 3 is probably better for you.

Go for Type 4 if you want the best quality and cutting-edge special features (like batteries, advanced monitoring, micro-inverters or hot water diversion), an in-depth consultative sale and a really well-planned install. Ditto if you have a difficult roof or a sensitive home (like mine – it's made of straw!). You should get a diligent installer who has the time to do everything just as you want.

Let's go through each type in more detail.

Type 1: Crap solar

In my opinion, a company that sells crap solar has the wrong values. Because they have the wrong values, their only point of difference is price. They're in a race to the bottom.

This infects their whole culture. They buy crap hardware if it is cheaper, they're relentless in trying to sell you their systems (you'll

Step 6: Getting Quotes

get pestered on the phone to buy), and their installs are unlikely to be done with the care and rigor that a high-voltage electrical system on your family's roof deserves. Their customer-unfriendly warranties often breach Australian Consumer Law, but they get away with it because most consumers don't know their rights.

Guess what? I don't recommend these companies, and I do everything I can to avoid working with them.

Type 2: Cheap but decent solar

When I first got started in the solar game almost ten years ago, there were three price levels for solar-power systems: cheap, mid-range and expensive.

To get a decent system, well installed with good after-sales support, I strongly counselled people looking for good value to go mid-range. The cheap stuff inevitably cut corners, and the companies generally shirked their responsibility to support you if things went wrong down the track.

As the solar industry has matured, I've seen good companies come on to the scene selling cheap (but not the absolute cheapest) systems that are well installed, using reputable, big-brand panels and inverters, and supported if anything goes wrong.

This new breed of cheap solar company offers cheaper-than-average pricing by being ruthlessly efficient in their operations and offering a no-frills but professional service.

They're not for you if you want a long conversation with the installer or designer, lots of options to choose from, a complicated install, advanced features or plenty of hand-holding to get the most out of your system once it is up and running.

This type of company could work for you if:

- your house is a simple layout
- you live in a large metro area
- you want a decent solar-power system of a standard 3, 5, 6 or 10 kW size, and
- you're happy with a Jetstar-style experience over Qantas

The trick is to differentiate the good-value, cheaper-than-average guys from the bottom-of-the-barrel, absolute cheapest companies that are likely to disappoint.

Here's how, in my experience, a reputable company can sell at lower-than-average prices:

1. They have really efficient operations. The larger companies may achieve this through investing in an IT system and business processes that make sure their processes are repeatable, efficient and consistent. Or smallish companies with low overheads may have been doing this thing for ages and run like a well-oiled machine.
2. They often import their solar panels directly from the overseas manufacturer by the container load instead of through the manufacturer's Australian office. This is legitimate – as long as they're importing the panels approved for use in the Australian market and they're not 'grey' imports.[14] To be safe, these panels must be a big brand name, like the ones mentioned in the last chapter.
3. The disadvantage is that you only have one point of contact for your warranty, because the importer is also the solar company. If the solar company goes under, it will be much

---

[14] Learn about grey imports here: solarquotes.com.au/grey.

Step 6: Getting Quotes

      harder to claim on the panel warranty than if they were imported by an independent, third-party entity.
4. They quote over the internet using satellite imagery of your home. This saves on the cost of an initial home visit.
5. Their prices are based on simple jobs. Usually, that's up to two roof areas, without any shading issues, on a conventional roof.
6. You don't get much choice of solar panels or inverters. Usually, they offer only one or two brands.

Expect a professional – but not time-consuming – personal service. Their business model does not allow for lots of hand-holding.

At this pricing level, look for companies that have been around for years. This shows that their overheads are sustainable. New, cheap solar companies often hit the wall when they realise their margins don't cover their real-world overheads.

Type 3: Happy medium

These guys set their prices at a level that allows them to run a sustainable business and still offer the following:

- A home visit and a discussion of your exact requirements.
- Simple consumption monitoring and on-site shade analysis to get a good estimate of your savings.
- A choice of hardware.
- Consideration of any special features, including:
  - advanced monitoring
  - hot water PV diversion
  - batteries – now or in the future, and
  - tricky or sensitive installs

Basically, this type of installer is more Qantas, less Jetstar.

Type 4: Expensive (craftsperson)

These guys are real craftspeople, and their installs should be impeccable, with every attention to detail. They only work with the best and most expensive brands (often only LG or Sunpower panels).

If you're thinking of storing electricity, they're usually all over battery technology, standards and configuration. They'll be happy to come back to their installs many times to make sure the battery-control software is fine-tuned to your consumption and the seasons, maximising your battery savings.

If you want a really special system with the best hardware, killer performance per square meter of roof, bespoke features and an install that could sit in an art gallery, then these guys are a first-class choice.

Type 5: Expensive (rip-off)

These guys tend to have a business model that's based on you only getting one quote – so you can't see how expensive they are for what you're getting.

They often employ the industry's most persuasive salespeople, and they prefer exclusive appointments – through either door-knocking or annoying unsolicited phone calls from their pushy (usually offshore) appointment setters.

Expect the hard sell from these mobs – and know your rights.

For these guys, maximising the sales commission is the name of the game. The higher the price they agree with you, the bigger the commission. To make matters worse, they often sell mid-

Step 6: Getting Quotes

range or even really cheap, no-name hardware to make those margins even fatter.

Avoid these guys. Especially the door-knockers.

How do you find the good guys? Research is the key.

How to find a reputable solar installer (or three)

As you learned way back in Step 1, the solar industry is driven by the STC scheme, which everyone but the government and the regulators calls the solar rebate.

The rebate covers almost half the price you pay for solar, and the government uses it as a way to regulate the industry.

In a nutshell, if you want to claim the rebate, you have to follow some extra rules and regulations when installing a solar system.

You have to use approved inverters, approved panels and an approved installer. The panels, inverters and installers are approved by a non-profit called the Clean Energy Council.

All companies that sell grid-connect solar systems must use approved hardware. They must also use accredited designers to design the system. To install the system, they must use licensed electricians who have a further qualification that classes them as a Clean Energy Council Accredited Installer.

> **Key point:** The Clean Energy Council only accredits people, not companies.

Some solar installation companies claim to be Clean Energy Council-accredited. This is not true. The Clean Energy Council only accredits individual electricians. By law, companies need to use those accredited electricians to install residential solar under the solar rebate (STC) scheme.

While a company can't be accredited by the Clean Energy Council, it can be a member of the organisation. Unfortunately, being a member isn't a watertight indicator that a company is reputable. The Clean Energy Council survives on subscription fees, so most companies that apply to be a member will be approved – unless they've done something really bad. Scrolling through their member list as research for this book revealed a couple of companies that I wouldn't recommend to anyone.

However, the Clean Energy Council does have a scheme that has a high bar to entry and excellent compliance checks: the Approved Solar Retailer Scheme.

Clean Energy Council-Approved Solar Retailers

If you're looking to engage an installer and they're an Approved Solar Retailer (again – don't confuse this with a 'member' or an 'installer'), they're almost certainly a reputable company.

Clean Energy Council Approved Solar Retailer logo

Step 6: Getting Quotes

Does that mean the solution is to get a quote from the nearest Clean Energy Council-Approved Solar Retailer? If you have one locally then I would absolutely recommend getting a quote from them. You can find an up-to-date list of every approved retailer here: solarquotes.com.au/asr.

If you can find three companies using the link above who can install locally, getting a quote from each of them will give you a choice of quality systems backed by reputable installers.

However at the time of writing, there are only 54 Clean Energy Council-approved solar retailers across the entire country. There are thousands of companies selling solar that are not Clean Energy Council-Approved Solar Retailers, including dozens that I know to be absolutely excellent operators.

The downside of limiting yourself to Clean Energy Council-Approved Solar Retailers is that with a little over 2% of solar companies enrolled, the chances are that you'll be excluding many of your local companies and you'll have to go further afield for quotes. The price you'll pay may be a little higher, and the choice of hardware is likely to be narrower, than it could be if you used a well-regarded local.

Why aren't more solar companies enrolled in the Clean Energy Council Approved Solar Retailer scheme?

- **The cost.** Depending on the size of the company, being a Clean Energy Council-Approved Solar Retailer costs $800 to $6,000 a year in fees – plus higher overheads due to a strict compliance requirement for procedures and policies.
- **Its requirements favour medium to large organisations.** Small installers have told me they don't feel the scheme is

suited to small installers with very low overheads who value autonomy and dislike bureaucracy.
- **Approval is a high bar.** It takes a big effort to comply and stay compliant. Small solar companies don't have a whole lot of time or resources to deal with this stuff.
- **Politics.** Without getting into the politics of the Australian solar industry, many solar companies are wary of the Clean Energy Council and disagree with some of their procedures and policies. I like most of what the council does, but I respect solar installers' grievances.

There are two other things to bear in mind about the scheme:

- **If you have a complaint about an approved retailer, the Clean Energy Council will investigate.** They'll kick the company out of the scheme if they've breached the code of conduct, but they won't help you resolve your particular issue. They'll simply direct you to your local consumer tribunal.
- **Being an Approved Solar Retailer isn't a guarantee of the company's financial stability.** Two approved companies have gone broke. As I mentioned in Step 5, though, it's impossible to guarantee a company won't go broke, especially in the low-margin, high-cash-flow solar installation business.

The Smart Energy Council's Master Installer badge

The Smart Energy Council[15] is a more 'grassroots' version of the Clean Energy Council. Its fees are lower, and its members tend to represent the 'small end of town' better.

---

[15] Called the Australian Solar Council until late 2017.

Step 6: Getting Quotes

They have a Master Installer Scheme, where installers can pay a fee, do some extra training and use the moniker 'Master Installer'.

At the time of writing, about 130 installers are signed up, but the scheme is less transparent than that of the Clean Energy Council about how installers qualify. From what I've seen, the bar is quite low compared with the Clean Energy Council scheme.

> **Online resource:** You can get quotes from Smart Energy Council Master Installers here: solarquotes.com.au/asc

If you can't find three local Clean Energy Council-Approved Solar Retailers for your quotes, it's certainly better to look for local Smart Energy Council Master Installers than to do a Google search. An installer who makes the effort to become a Master Installer has shown some commitment to the industry.

Other ways to find installers

I'm a big fan of going local when buying solar. There are almost five thousand qualified solar installers in Australia, so no matter where you live, there is likely to be at least one close to you. With any luck, there will be at least three, so you can do the sensible thing and get three quotes to compare.

Manual search. Here's how to find the solar installers nearest to you:

1. Go to the official Clean Energy Council 'find an installer' page: solarquotes.com.au/findcec.
2. Click on 'search by location' and enter your suburb.
3. The map will refresh with a red pin for every local electrician who is qualified to install solar. Click on a pin for the details. It will give their individual name, company

name (unless they are a sole trader) and phone number.

This is a great way to find your nearest solar-qualified electrician. The only problem is determining how good they are at solar installs and the quality of the hardware they choose to use.

If you put the electrician's name and 'solar' into Google, you may find some reviews that will give you an indication of their quality. I'd also recommend searching for their name on my site (solarquotes.com.au), which has over 25,000 reviews of solar installers.

If you luck out and find three local electricians who install solar themselves or work for a well-reviewed solar installation company, I recommend that you ask those three to provide quotes for the system size you need.

Using solar quoting services or brokers. These days, if you go online and put 'quotes for solar' into Google, the results will be dominated by companies offering to get you quotes from multiple companies for free.

These websites have a client base of solar installation companies around the country. When you fill in a form on the website to ask for quotes, they refer you to several companies and charge each company a referral fee. Some also take a commission from the solar companies if you choose to go with one of them.

I own and operate one such company: it's called SolarQuotes®. Since 2009, over 1 in 30 homes in Australia has used it to get quotes.

Over the last few years, lots of copycat sites have sprung up, but from what I can see (and, yes, I am biased), most of them fail in one big way: they don't vet the solar companies that they

Step 6: Getting Quotes

recommend very well. Many of these sites have clients that have been fined or reprimanded by the ACCC for false claims or fake reviews, so be careful.

There are only two solar quoting websites that I would recommend. Mine: SolarQuotes® (solarquotes.com.au) and Solar Choice (solarchoice.net.au).

The main difference between the two sites is that SolarQuotes® hosts tens of thousands of reviews of almost every solar company in Australia and is mostly focused on residential solar, whereas Solar Choice does not host customer reviews, and, in my opinion, is a better choice for large, commercial solar projects. The other thing you should know is that SolarQuotes® only charges a referral fee (about $40 per referral) to the company. Last time I checked, Solar Choice charges a referral fee plus a commission on any sale made to you.

SolarQuotes® and Solar Choice both vet their installers carefully, so using their services is a low-risk way to quickly get up to three quotes from reputable solar companies.

I would recommend using either one or the other. If you use both, you're likely to get six quotes, which can become overwhelming for everyone involved.

How not to buy solar

Finally, here are some common mistakes to avoid.

Don't get quotes from unsolicited callers or door-knockers. In my experience, companies that use these tactics almost always either charge expensive prices or use junk hardware. Or both.

Don't be tempted to buy second-hand solar equipment. I'm a big fan of second-hand cars, bikes and furniture, but not solar. I won't go into the reasons why here, but you can read about the specific problems on my blog: solarquotes.com.au/secondhand.

Don't buy solar on eBay or Gumtree. Buying any solar equipment – apart from little camping panels – on eBay, Gumtree or similar sites is fraught with danger. If the panels are sent from overseas, they're unlikely to meet the strict Australian electrical safety standards and it's likely to be illegal to connect them to the grid. You're also unlikely to save any money, because you can't claim the rebate on unapproved equipment.

Don't look for the absolute cheapest system on the market. If a solar price seems too good to be true... well, you know how it goes. I talk about the specific dangers of the cheapest solar on the market on my blog – just in case my earlier rants didn't convince you: solarquotes.com.au/cheap.

Site inspection

When it comes to getting firm quotes, I strongly recommend that you have a physical site inspection.

Since I started in the solar industry in 2009, the most frequent complaint I've had from people looking for quotes is: 'This solar installer won't even give me a ballpark price for a solar system! He's insisting on coming round to my house first!'

I understand the sentiment. If you want *really* ballpark figures, I maintain a list of the approximate cost of good-quality solar systems here: solarquotes.com.au/cost.

Step 6: Getting Quotes

Just be aware that those prices can vary by thousands of dollars depending on your home's layout and the condition of your roof. If the prices on that page don't have you running for the hills then the next step, in all honesty, is to have each solar installer or salesperson come round and assess how suitable your home is for solar power.

A valid reason for resisting a site inspection is that you've heard reports of some solar salespeople pressuring people into buying. I believe this is completely unacceptable. If anyone selling solar tries to get you to sign anything without giving you the courtesy of going away so you can consider your options, my advice is to decline their offer.

Oh, and while we're on the subject, if you ever come across one of the special breed of jerks who try to get you to waive your cooling-off period, please kick them out of the door first, report them to ACCC second, and, third, let me know so I can put them on my blacklist.

The reality is that most solar salespeople and solar installers are really nice people who just want to come round so they can be sure that they recommend a system that suits your electricity usage patterns and your home. Of course, they also want the opportunity to impress on you why their solution is the best, but that's all part of the fun.

Here are the reasons I recommend a site inspection:

- **You can look the installer or salesperson in the eye.**
  Assess their competence by asking some tough questions. If they can't answer them or they look panicked, end the visit promptly and move on.

- **Your roof needs checking.** The installers need to check the condition of your roof if it's more than ten years old. It may need repairs before anything is installed. You really don't want to have to remove your panels to repair your roof a few months or years down the line.
- **Shade needs assessing.** If you have any shade on your roof at all, an installer can't assess how much that will hurt your solar-power system using Google Earth.
- **Cables need routing.** The installers need to assess how they're going to get the cables from the solar panels to the inverter with as little modification to your house as possible.
- **Switchboard may need upgrading.** They need to look at your switchboard to see if it needs upgrading and if there is space inside it for an extra circuit breaker.
- **Inverter location needs deciding.** The installer needs to scope out a place for the inverter where it is nicely shaded, out of harm's way, and as close to the meter as possible.
- **Panel locations need reviewing.** If you're getting a big system (at least 5 kW), roof space is likely to be at a premium. You should really get a more accurate measurement than Google Maps can provide. Also, you may need to split your solar panels over multiple roof areas. If this is the case, you're much more likely to get a better designed, better performing system if the designer can visit your home.
- **Future battery location needs considering.** Most solar homes in Australia will add batteries at some point. Pre-empting where they might go and finding an appropriate place for the solar inverter makes a lot of sense if you have one eye on the future.

Step 6: Getting Quotes

- **Energy consumption needs measuring.** Having a look at your home and appliances and chatting about your lifestyle will give a good solar installer a great feel for your self-consumption ratio (if you haven't measured it) and opportunities to improve it, which will make your solar payback even better.

Having said all that, you may want a small system, have no interest in getting batteries in the future, and have a recently built home with a massive north-facing roof, a modern switchboard and no trees or other shading issues whatsoever. In that case, you could argue that buying solar without an inspection is pretty low risk – and it is.

Even if you're only paying $4,500 for a small 3 kW system, though, don't forget that with the rebate, it's actually closer to $7,000 worth of gear that you'll use every day for the next 20-plus years. An inspection will hugely increase the chance that you get a system that performs well and makes you happy. It's kind of a no-brainer, isn't it?

A comprehensive quote

A sign of a professional installer or installation company is a comprehensive quote. Here is the information that you can expect to see on each of your quotes:

Item 1: An itemised list of the hardware to be supplied

At a minimum, this should include the following:

- **Solar inverters:** Quantity, brand, model, AC rating (kW), number of MPPTs and warranty, including warranty extension options.

- **Solar panels:** Quantity, brand, model number, panel size (W), total array size (kW), product warranty and performance warranty in years.
- **Racking:** Brand. (Good brands include Radiant, Grace, Sun Lock, Clenergy and Schletter.)
- **Monitoring system (if needed):** Brand and any ongoing subscription fees.
- **Extra charges:** split arrays, two-storey roof, switchboard upgrade, tilt frames etc.

Item 2: Installation warranty

The installation warranty covers defects in the installed system that are not specific to the panels or inverters, for example, if a cable becomes disconnected, an isolator fills with rain and fails, or the racking system comes loose.

- Don't accept less than 5 years
- If the installation is subcontracted, don't accept a clause that places responsibility for honouring this warranty with the subcontractor. That's not allowed under Australian Consumer Law.

Item 3: Total price of all goods and services

This includes the following:

- The gross cost of the system including GST
- Discount from STCs (AKA the solar rebate)
- Any extra charges such as for two storeys, tiled roofs or split arrays
- Any state-based rebates

Step 6: Getting Quotes

Here's an example:

Your price

System		4.16kW
Panels		16xTrina 260W 'Honey' Module
Standard Inverter		SolaX X1 SL-TL4400T (Dual MPPT)
Price before your cashback on Small-scale Technology Certificates (STCs)		$10,687.00
STCs	x 86	
STC price	$39.50 per certificate	-$3,397.00
Price after your STC cashback	Standard installation	$7,290.00
Split-array (5 arrays)		$864.00
Your price (includes GST)		**$8,154.00**

Figure 6.1 Price section of a quote.

Key point: STC prices change. We learned in Step 1 that STCs can change in value every day. Your quote must clearly state whether the quote has fixed the STC price or not. For example, if the market STC price drops between now and the day of the installation, that reduces the value of the rebate.

In the example above, a drop of $3 per STC would reduce the rebate value by 86 STCs x $3 = $258.

Will the installer wear that cost or increase your price?

This must be clearly disclosed – not hidden in the small print.

Item 4: A sketch or diagram of the proposed roof plan

This should include an itemised list of every panel array's direction and tilt.

For example:

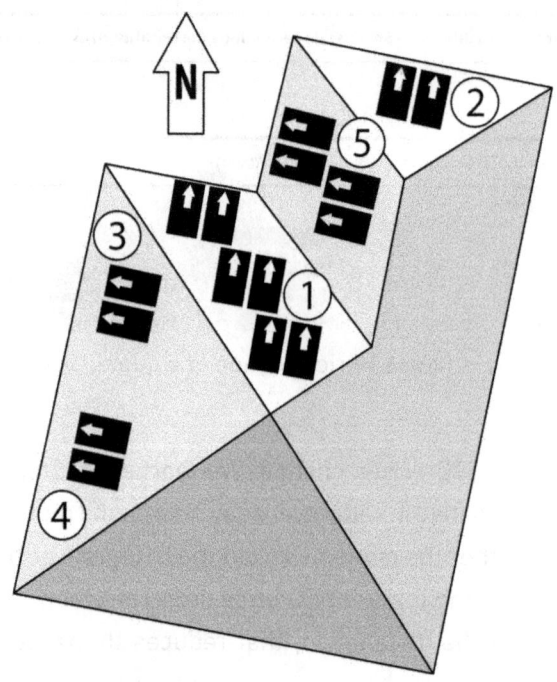

Figure 6.2 Roof plan showing five separate arrays to be installed and an existing solar water heater.

Panel arrays:

1. Six panels: direction 10 degrees clockwise of north, tilt 30 degrees from horizontal
2. Two panels: direction 10 degrees clockwise of north, tilt 30 degrees from horizontal

Step 6: Getting Quotes

3. Two panels: 10 degrees cloc kwise of west, tilt 30 degrees from horizontal
4. Two panels: 10 degrees cloc kwise of west, tilt 30 degrees from horizontal
5. Four panels: 10 degrees cloc kwise of west, tilt 30 degrees from horizontal

It is important to itemise each separate panel array because it shows that the installer has accounted for the expense of cable penetrations through the roof at every array. An install done on the cheap might only have one roof penetration, with the arrays connected with ugly electrical conduit going all over your roof like spaghetti.

Item 5: Your expected efficiency losses due to shading

If your roof has any shade at all, the quote should include how the installer has calculated these losses. This post explains how I believe it should be calculated: solarquotes.com.au/shade.

Item 6: Estimated energy yield for your roof

I like to see a chart or table with estimates of the system's average daily performance (in kWh) for each month of the year.

Figure 6.3 is an example from a Queensland quote.

A chart like this on your quote serves two purposes. Firstly, it shows that the company quoting is a true pro, because they've invested in a system that accurately models system performance. Secondly, it lets you assess if you'll have enough energy through winter. You can compare the worst months with your measured daytime usage, with extra kWh added on for water heating if that applies to you.

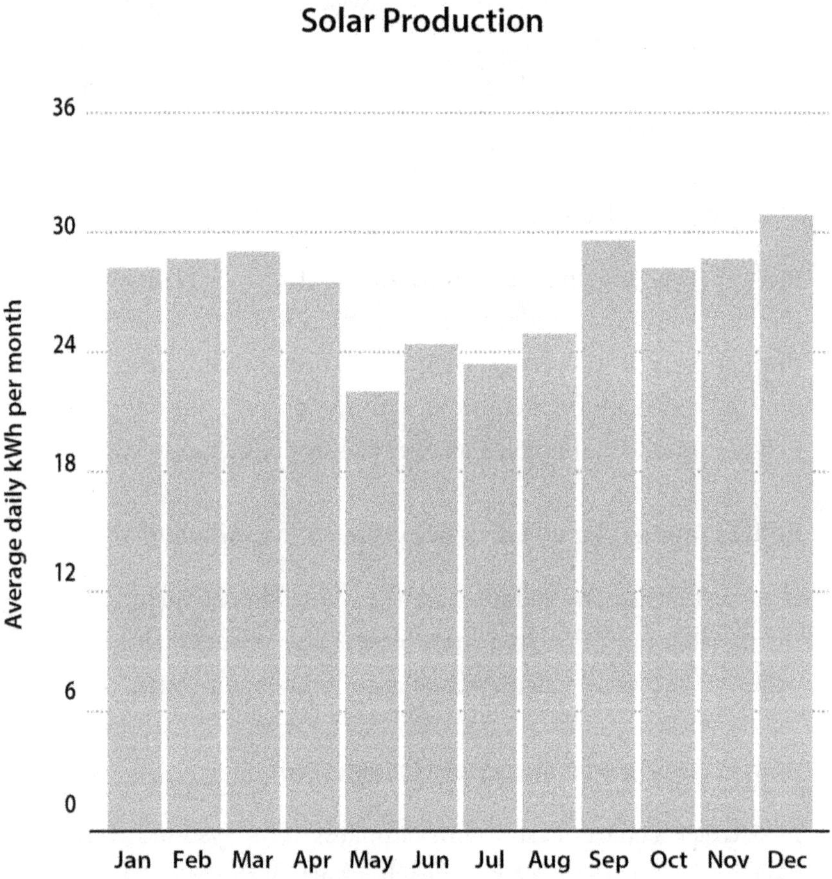

Figure 6.3 How a good quote presents expected solar production for a Queensland system.

Monthly variation isn't a big deal for lucky Queenslanders. But if you live in South Australia, you really need to consider it.

Here's a chart from an Adelaide quote:

Step 6: Getting Quotes

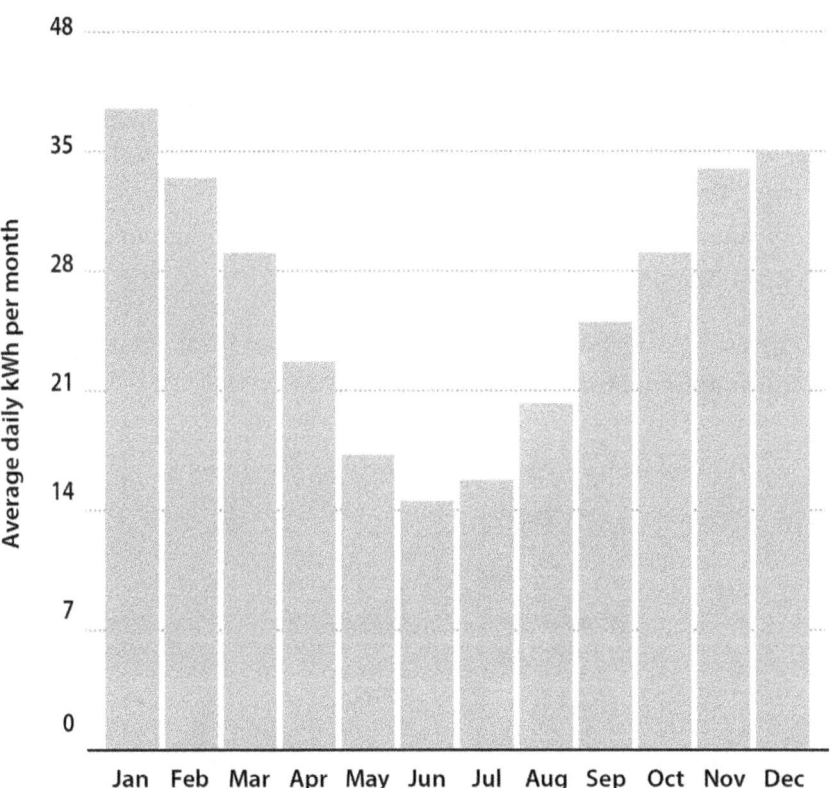

Figure 6.4 Variable monthly production in Adelaide.

Item 7: Savings projections and associated assumptions

Any solar seller who has confidence in the systems they're selling should be happy to put your estimated savings in writing. Your actual post-solar bills will depend on how much energy you use after the system is installed. If you throw caution to the wind and use twice the electricity after the install, your bill may not improve or may even get worse. If you keep your consumption the same, it's easy for an installer to predict your savings.

Self-consumption ratio: As we've learned, the biggest thing affecting projected savings is your self-consumption ratio. That means it's important for any projected savings to show the self-consumption ratio they've assumed. Check this is close to the one you calculated in Step 2. If the ratio is based on 100% self-consumption, that's a red flag. I don't know any household that consumes 100% of their solar.

Figure 6.5 shows how a good quote may present your self-consumption ratio – so you can sanity-check it:

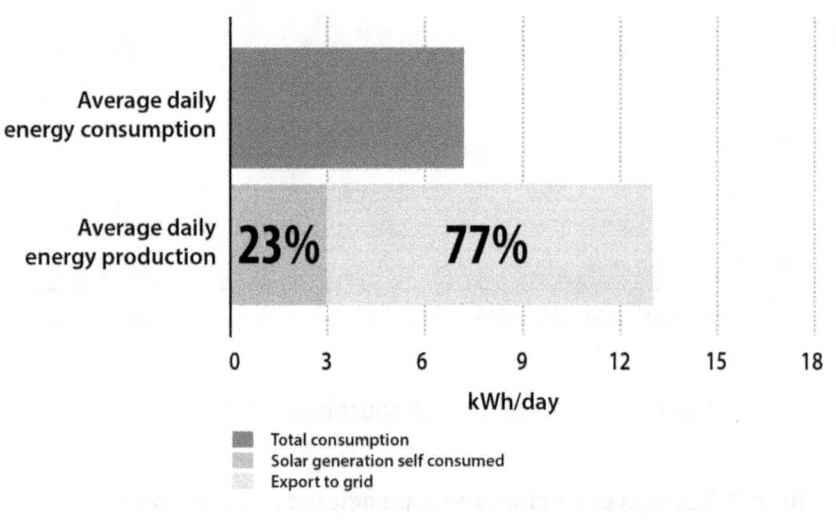

Figure 6.5 How a good quote presents the self-consumption used in savings estimates.

As you can see, this customer is predicted to self-consume 23% of their solar generation.

Step 6: Getting Quotes

Internal rate of return: The quote may also calculate your internal rate of return based on a fixed timeframe. Twenty-five years is reasonable if the panels are warranted for that amount of time.

One thing to bear in mind is inverter replacement. No inverter is likely to last 25 years, so the internal rate of return should include the cost of one replacement inverter. Check the installer has included this.

Total savings: If a modest annual electricity price increase is assumed, the total savings over 25 years can be quite astonishing.

Figure 6.6 demonstrates a good way to present total savings. Be sure that the quote shows the inflation of electricity prices they have assumed and make sure you agree with it.

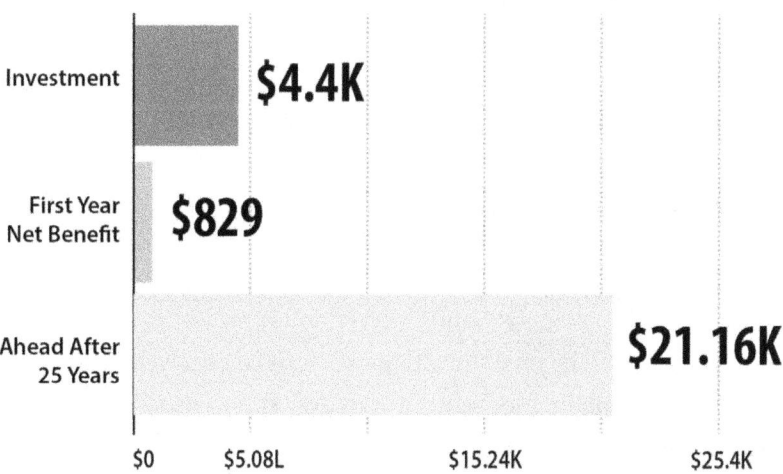

Figure 6.6 Return on investment over 25 years.

Bill before and after solar: This is the most important projection for most people. What can you expect your post-solar bills to be? After you get solar, you're unlikely to maintain a spreadsheet and track your internal rates of return, but you will get a bill every quarter – and if it's low enough, you'll be happy.

Figure 6.7 is an example of a good way to present the difference between your bills 'before' and 'after'.

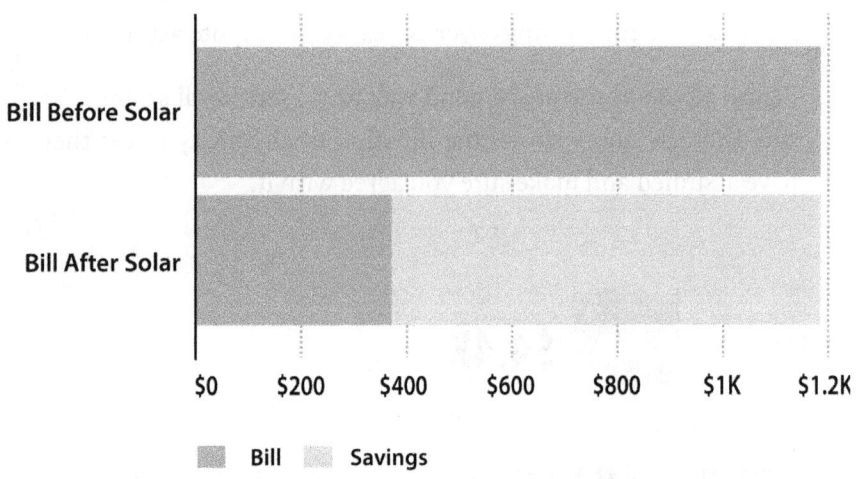

Figure 6.7 A good quote shows your projected bill reduction.

Energy graph: The final thing that's nice to see on a well-presented quote is a graph of your solar production and usage. They'll need to have estimated or measured your hourly energy-use profile to create this graph, and it gives you a good feel for how the solar will work with your usage.

Step 6: Getting Quotes

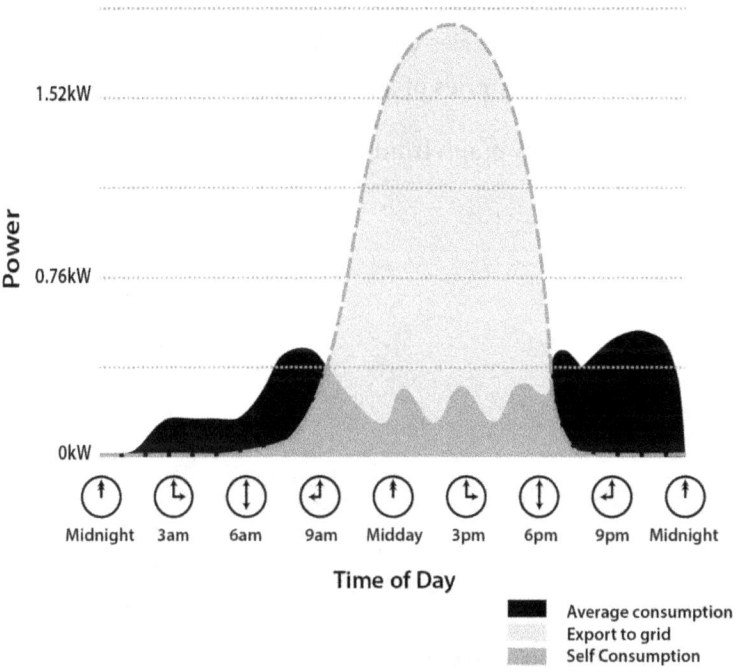

Figure 6.8 Projected solar production and household consumption over a winter's day.

On this graph (for a winter's day), we can see how the solar system easily covers the average load of the house. It will cope with small spikes in demand and have plenty of energy to charge a battery in the future.

The black area shows grid imports; the light-grey area shows grid exports. Because the light-grey area is bigger than the black area, you can be confident that this system will provide enough energy on a typical winter's day to charge a battery that will cover your night-time consumption.

If you're buying batteries, an energy graph like this is essential to show you how your batteries will work, so you can be confident that you are getting batteries of an appropriate size.

Figure 6.9 is an energy graph from a quote for solar and batteries.

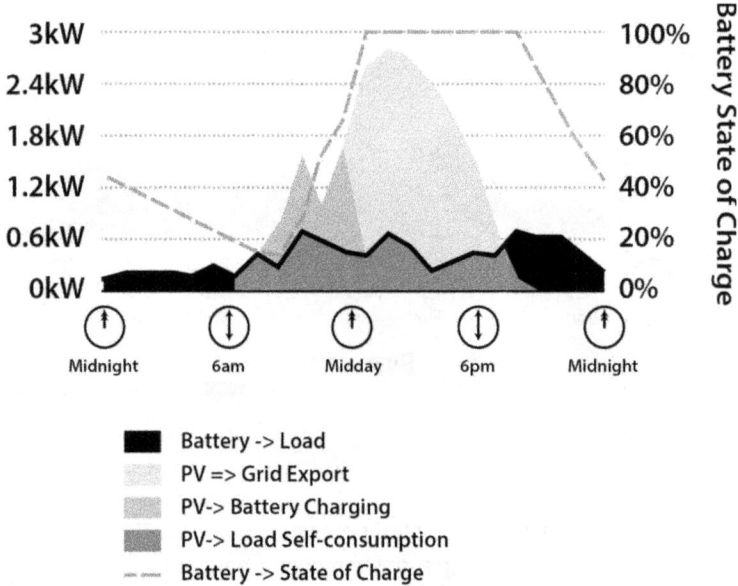

Figure 6.9 A day's projected generation and consumption for solar and batteries.

This graph assumes a well-sized battery. It charges in the morning (pale grey) by midday in summer – which means it should charge fully through winter. The black sections show that the battery powers the home through the night. The dashed line shows that you use most of the battery capacity, as it discharges to about 10% in the morning. That means you haven't wasted money on battery capacity you won't use.

Step 6: Getting Quotes

Terms and conditions

This would be a thick book if I were to go through all the possible legal terms in a contract and explain what they mean, so I'll just give you the 'gotchas' to look out for:

- Check that the terms are written in a clear and transparent way, using plain language. If they aren't, go elsewhere. In twenty-first century Australia, there's no place for terms that are written in legalese. This is a contract for thousands of dollars of your money and, at the very least, the terms should be clearly written.
- Look for clauses that lay out any circumstances where you may have to pay above the agreed price. For example, do you have to pay extra for meter replacement or reconfiguration? If you haven't had a site inspection, this is really important – it's highly likely that when the installers rock up, they'll find things that create work for them above what the sales guy on the phone has accounted for (for example, switchboard modifications).
- Check how you have to pay, how much the deposit is (don't pay more than 10%) and when the balance will be payable.
- Check your rights to end the agreement if the installation is seriously delayed.
- The installer will usually have to do a more detailed design once the contract is signed. Check that you can get a refund of your deposit if the final system design they provide is significantly different from the one quoted when you signed the contract.
- Check what happens if the installer can't get permission from your DNSP (your electricity network) to connect the

The Good Solar Guide

system to the grid. This needs to be done before any work starts – and if they don't get permission, you should be able to get your deposit back while you consider your options.

- Check for a clause that says the installer can switch the agreed brands of panels or inverters for 'equivalent' ones if they're out of stock. If you find a clause like this, do not sign. It's a trick used by cowboys to quote for good brands and install junk.
- Check for a clause that insists on an 'annual inspection' to maintain your warranties. This is BS and violates your rights under Australian Consumer Law.
- Finally, look for a clause similar to this in the small print: '[The company] accepts no responsibility whatsoever, in the event that the performance of the PV System is lower than predicted.' I pulled that from ACTEW AGL's contract for solar sales. Don't stand for it. The company should do everything they can to help you if the performance of the system is substantially different from what they promised.

A good installer or salesperson will talk you thorough the contract and carefully highlight any terms that may result in delays or extra costs. If they make any assurances or promises that are not openly stated in the written contract, ask them to write them on the contract. All verbal promises are 'express warranties' under Australian Consumer Law and the company has to honour them.

Once you've chosen a quote, only sign it when you're 100% sure about what you're getting, when it will be installed, and the how long it will take to have your meter changed over. You won't be able to switch on your system until that last step happens.

Step 6: Getting Quotes

If, at any point, you feel pressured into signing something – don't. A good installer will not pressure you to sign their contract.

Summary

- Familiarise yourself with the five types of installers. Avoid the cheap crap and the super-expensive rip-offs.
- Avoid unsolicited door-knockers at all costs.
- Getting three quotes is a good way to protect yourself.
- The Clean Energy Council's Approved Solar Retailer Scheme is different from its Accredited Installer scheme. In my opinion, the Approved Solar Retailer scheme is excellent and takes most of the risk out of buying solar. You're highly unlikely to go with a bad 'un if you choose a company on the approved retailer list, and you'll almost certainly end up with a well-installed, quality system. Just keep in mind that because of the low enrolment rate, you're likely to pay a little more, have a narrower choice, and have to go a little further afield than if you choose to include installers who aren't on the Clean Energy Council-Approved Solar Retailer list.
- The Smart Energy Council's Master Installer scheme is a lower bar – but it's still a good sign that the installer cares about their industry.
- There are lots of 'get three quotes for solar' websites. Don't get the many copycat sites confused with the only two I recommend: solarquotes.com.au (which I own) and solarchoice.net.au (the only other site I would advise a friend to look at).
- I strongly recommend using installers who will do a site inspection as part of their quote.

- A comprehensive quote with detailed savings predictions and transparent assumptions is a sign of a professional solar company.
- Always look at the terms and conditions attached to the quote.

STEP 7: After The Install

Your solar system has been installed. What next?

Unless your solar installer is also qualified to work on your meter (usually, they're not), you'll have to wait for a separate sparky from your electricity retailer or network to come and reconfigure your meter to measure your exports. They may even need to give you a whole new meter.

If it isn't part of your installation quote, this usually costs about $300. It'll cost more if you have a three phase supply.

You are not normally allowed to switch your solar system on until your new meter is installed. To help pass the time, you can do some simple checks of the install. I'm not recommending that you climb on the roof or try to do the job of a trained inspector, but there are simple things that can get overlooked. You can easily look for these and quickly get them put right so your system has the best chance possible of lasting 25 years or more.

Checking your install

Here are the things to look for.

On your wall

Check the following things:

1. Inverters and isolators are protected from lengthy exposure to direct sunlight.
2. Any outside, wall-mounted isolators do not have any electrical conduit entries from the top. This is a recipe for rain getting into the isolator, and that's big trouble.
3. The cables going into the inverter are neat and secured so they can't be pulled or caught.
4. All open conduits are sealed with a gland, not silicone.
5. The inverter (or third-party monitoring system) is set to alert you if something goes wrong, so you don't have to wait for the next bill to find out.

On your roof

If your house is single storey, you should be able to see all this from the ground – don't go climbing up there unless you know what you're doing.

1. The panels are neatly lined up and level. Anything else is just plain lazy.
2. Any excess rails are trimmed.
3. No panels are overhanging or less than 200 mm from the edge of the roof. Positioning panels close to the edge is asking for the wind to get under and damage your roof or stress your panels. If the panels are too close to the gutter, the rain running off them can miss the gutter altogether.

STEP 7: After The Install

Online Resource: If you really need to go closer than 200mm to the edge of your roof - it is possible with the right racking and enough roof fixings. This post explains how: solarquotes.com.au/edge

4. Any cables between panel arrays are not via an ugly, sun-exposed electrical conduit but through the roof cavity.
5. The solar panel clamps are in 'clamping zones'. The clamping zones vary between panel makes and models. Usually, each panel will have four clamps on the long sides and they need to be 100–300 mm from each corner. If the clamps are too far away from or too close to the corners, your panel can flex in the wind, cracking the silicon cells and drastically shortening its life.

Online resource: Learn more about solar panel clamping zones here: solarquotes.com.au/cz

6. If your quote and design assumed no shading, your panels should be unshaded throughout the day. Be wary of TV aerials, roof vents and flues.
7. All rooftop isolators should be shielded from the sun.
8. Any tilt-racking legs should usually be at right angles to the panels for maximum strength. If they aren't, check that this is allowed by asking to see the racking manufacturer's instructions, which will contain the relevant diagram. I've noticed that solar installers are excellent at electrics but they can miss a trick on the mechanical side of things, so this is worth checking.

Online resource: You can find an illustrated checklist with photos of good and bad installations for each point here: solarquotes.com.au/installcheck

Documentation

Sadly, many installers – even the good ones – don't give documentation the priority it deserves, so you may have to hassle them for this.

Documentation is important because:

- you can refer to it if something goes wrong
- it shows you how to safely shut down and start up the system, and
- it tells you who to contact for warranties

If anyone comes to service or inspect your system, or do other electrical work on your house, it's an important reference for them too. Moreover, the Australian Standard for solar installation says it has to be provided, so insist on a full documentation package.

This should include the following:

- List of equipment
- Warranty information – including which manufacturers to contact if the solar company disappears
- Equipment manual
- Equipment handbook
- Array frame or racking engineering certificate (proves mechanical safety)
- Shut-down and isolation procedure
- System performance estimate
- Maintenance requirements
- What to do if there's an earth fault alarm
- System-connection diagram
- Site inspection checklist
- Testing and commissioning checklist
- Declaration of compliance
- Certificate of electrical safety

STEP 7: After The Install

Shifting loads

Once your solar is installed and your retailer has reconfigured or replaced your meter, the system will be powered on and it will start generating – hopefully, non-stop for 30 years (although you'll need to replace the inverter after about 15 years).

If you want to maximise your savings, now is a great time to think about shifting loads to the daytime.

From now on, when you use an appliance before the sun is up or after the sun's gone down, think about whether you could use it during the day instead.

Here are some examples to get you thinking.

Dishwasher: Morning

Do you get up in the morning, stack the dishwasher with breakfast plates, set it going and then leave the house?

Dishwashers can pull a couple of kW when they heat the water at the beginning of the cycle and again when they dry the dishes. (Anything that involves heating will use a lot of power.)

Your solar system's power will be fairly limited until about 10am, so why not use the delay-start feature so your dishwasher doesn't start until two hours after you leave the house?

Dishwasher: Evening

Whether you can pull off this trick depends on how big your dishwasher is compared with the size of your family. Is it possible to stack the dishwasher in the evening so you can still fit in the breakfast dishes? If yes, instead of running it overnight, you could wait until the morning and delay-start it.

Washing machine

Can you delay-start this in the morning? If possible, stagger it with the dishwasher – for example, put this on a four-hour delay and the dishwasher on a two-hour delay if you're leaving the house.

Clothes dryer

A conventional dryer is one of the most power-hungry devices in your home. It can chew though 3–8 kWh in one cycle.

If you have to use it, the middle of the day is best – but if the sun is shining, why aren't you sticking the washing on the line?

I would make it a priority to upgrade this appliance to a more efficient model. A six-star heat-pump dryer is up to four times more efficient than a regular dryer. In general, you'll only use the dryer when the weather isn't good enough for the clothesline, so you will usually have limited solar to power it. That means an efficient heat-pump dryer is a great upgrade for a solar-powered home.

Pool pumps

There's no good reason to run pool pumps at night. Run them as close to midday as you can, so most of their energy can come directly from the sun.

Hot water

If you have conventional electric hot water then you took my advice and got an automatic power diverter, right? If yes, you won't have to worry about when this switches on and off; the controller will make sure that it uses only solar whenever possible.

STEP 7: After The Install

If you have a good heat-pump hot water system, it should draw about 1 kW for two hours to get your water piping hot. Set its timer to start at 11am.

Heating and cooling

If you have a well-insulated home with plenty of 'thermal mass' (heavy materials, such as brick, stone and earth inside), you can pre-cool and pre-heat your home to take advantage of your solar power.

In summer, you can power on your air conditioner while you have good solar generation, and turn it down when the solar power drops. A good monitoring system will show you when you have enough solar. The thermal mass in your home will store the 'cool' and release it into your home through the evening.

In winter, you can do the same with heating.

If you have a brick veneer home or a weatherboard home, this won't be very effective. Your best bet is to upgrade to the most efficient reverse-cycle air conditioner you can afford. They go up to seven stars these days.

Checking your bills

Once you've had solar for a whole billing cycle, you should check your bill.

Look at the average daily usage shown on your bill. Remember, in terms of consumption, the bill only shows grid usage. It doesn't show how much solar your home has used (see earlier for the reasons why). If you use lots of your electricity during the day, you'll have a high self-consumption ratio and your grid imports should have dropped substantially.

Figure 7.1 shows an example of the average daily cost and usage before and after installing solar.

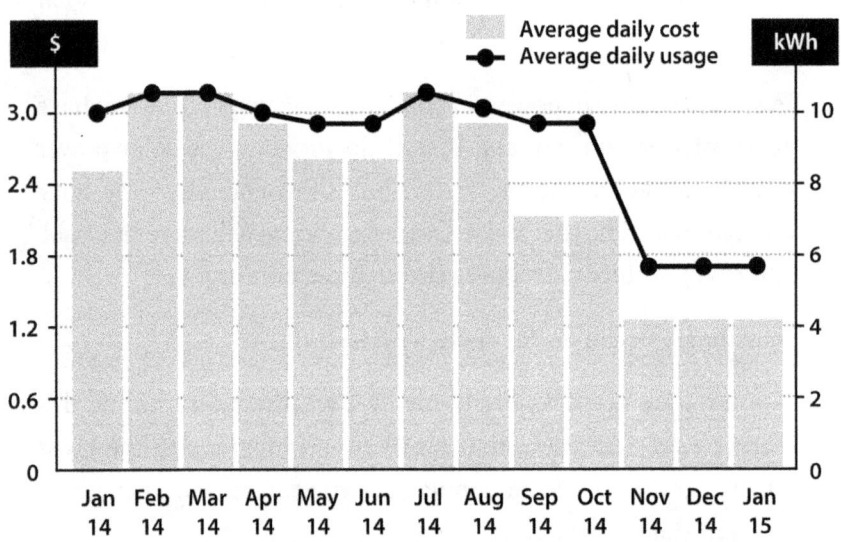

Figure 7.1 Extract from electricity bill for someone who installed solar early in November 2014.

In this example, the solar was installed at the start of November. The black line shows grid imports dropping from just under 10 kWh per day to just over 5 kWh per day. With a usage tariff of 30c per kWh, the daily cost has dropped by $1.50 per day.

Check what you measured your pre-solar daytime consumption to be. Your daily use should have dropped by about that much.

STEP 7: After The Install

If your grid consumption hasn't dropped since you got solar, there are three possible reasons:

1. There was a long delay between your solar system being installed and your solar meter being installed or configured. The solar system can only legally be switched on when the meter is ready.
2. Your solar system isn't working properly.
3. You have increased your evening and night-time consumption since getting solar.

If you took my advice and got solar monitoring installed, you'll be able to see what the issue is. The monitoring will show:

- the date your solar started producing
- if it is a good monitoring system, the health of your solar system (0%–100%), and
- how much electricity you're using – how much from the grid and how much from your solar

If you're confused or disappointed by your first bill after solar, call your installer and ask them to walk you through your bill and your monitoring. A good installer will be happy to show you exactly what's happening and how much money they're saving you.

Next, let's go through the charges on your bill.

Look at your 'solar buy back' or similarly worded line item (clue: look for the abbreviation 'CR').

On the bill shown in Figure 7.2, the solar owner gets two credits – one for their solar and one in the form of a discount as part of their plan. You can see that the owner exported 555 kWh at 8c per kWh. That's a total credit of $44.40.

New charges and credits

Usage and supply charges	Units	Price	Amount	
Peak	489kWh	$0.2274	$111.20	
Supply charge	28 days	$0.7162	$20.05	
Price change - 29 Jul 15 to 22 Sep 15 (56 days)				
Peak	614kWh	$0.2243	$137.72	
Peak next	366kWh	$0.22	$80.52	
Supply charge	56 days	$0.7162	$40.11	
Other charges				
Payment processing fee			$1.25	
Total charges				+$390.85
Credits				
Solar Buyback*	555kWh	$0.08	$44.40cr	←
4% Guaranteed Discount			$13.18cr	
Total credits				-$57.58cr
Total new charges and credits				= $333.27
Total GST				+$37.76
Total due (includes GST)				**= $371.03**
Discounted amount if paid by due date (includes $29.65 Pay on Time Discount) (includes $2.97 GST credit)				**=$338.41**

Figure 7.2 Solar export credits on a typical bill.

If you can't find a credit for your solar then call your electricity retailer to find out what has happened to it. It's not that unusual for your retailer to forget to credit you for your solar, so make sure you're being paid.

Also make sure that the rate per kWh is the one that was promised.

See how many kWhs you've been credited for. Divide it by the number of days billed. If it's a quarterly bill, that's 90 days. In this

STEP 7: After The Install

example, that's 555 kWh divided by 90 days, which comes to 6.2 kWh exported per day.

Check that this number tallies with the exports reported by your solar monitoring.

> **Online resource:** Are you getting a good deal? Here's a link to a blog post showing you how to compare your local solar-friendly energy tariffs so you can check: solarquotes.com.au/retailers

> **Tip**
> Your electricity retailer may try to put you on a 'Time-of-Use' (ToU) tariff before or after you install solar. They can't force you, so don't do it. This post explains what ToU is and why it is bad for solar owners: solarquotes.com.au/tou

Solar system maintenance

If your panels are tilted more than 10 degrees from horizontal, they'll self-clean in the rain. It's really not worth paying to get your solar panels cleaned unless some event has happened to make them spectacularly dirty.

> **Online resource:** It is counter-intuitive that most panels do not need regular cleaning. But here's a post explaining why it is usually a waste of money: solarquotes.com.au/clean

How often should you get your solar system inspected or maintained? In the end, that comes down to how safe you really consider your rooftop solar to be.

How safe is rooftop solar?

If you ever drive or ride in a car of your own free will, it probably makes sense for you to consider rooftop solar safe.

At the start of 2017 around 17.2% of Australian households had rooftop solar. This made for a grand total of over 1.6 million systems. On average, each home has had solar for 4.8 years, which comes to a total of over 7.7 million years without a fatality – and that's not bad.

Safety issues to consider

Unfortunately, rooftop solar is not perfectly safe. Anything with live current running through it can be dangerous if damaged or defective, and solar systems are no exception. Fires have resulted from faults and, while most have been small, some have resulted in whole buildings being burned to bits.

While solar systems have no moving parts to wear out, the following problems can occur:

- Cable insulation and conduit deteriorating over time
- Faulty components failing
- Components filling with rain
- Corrosion
- Animals chewing on cables
- Damage from natural disasters, such as earthquakes, bush fires and storms
- Damage from home renovations
- Incompetent installation
- DC isolator fires

STEP 7: After The Install

Having a professional inspect a system can result in problems being identified and put right before they become a danger.

Inspections: How much and how often?

The going rate for an inspection by an accredited solar installer is typically $200 to $300. For that money, they'll check the connections, cables, panels, rooftop mounting, DC isolator switches and inverter. Some offer to do extra tests for more money as part of a premium service, but as far as safety is concerned, I don't think the expense is worth the benefit.

In South Australia, Canberra and Victoria, you have to have your system tested every five years by your DNSP. This has to be done by a qualified installer – so arrange for them to do a comprehensive inspection at the same time as their tests. Expect to pay around $200 for this.

I would suggest getting an inspection and system test every five years, even if your DNSP doesn't make you do it.

The final choice is down to you. If you want the sense of satisfaction that comes from knowing you've done something to keep your family safe, even if it's only a small thing, have your system inspected every five years to make sure it's in good working order.

Summary

- Check your install using the checklists at the beginning of this chapter. A good solar installer should happily put any issues right. Even better, before you engage the installer, agree with them that these things have to be done.

- Use your monitoring system (please tell me you got monitoring!) and common sense to get you and your family into the habit of using more power in the daytime and less at night.
- Once you've had solar for a full billing cycle, check your bill to ensure your grid usage has dropped and your feed-in tariff is getting paid.
- Use the tool I showed you (which uses the exact numbers on your bill for consumption and exports) to see if you can get a better deal on your electricity.
- Mark your calendar to get a safety inspection in five years' time.

CONCLUSION: Beyond Solar

You've followed the seven steps. You've done your homework and if you decided that a solar system is a good investment, you've engaged a good installer and filled your roof with solar panels. You've shifted your loads and found a solar-friendly electricity retailer, and your bills are nice and low. You know your system will alert you if there's a problem. You're starting to look forward to electricity bills rather than dreading them.

That's a great place to be – so what's next?

Efficiency

If you have solar, your bills should have dropped significantly. If you have an efficient home, you really should be looking at close to zero dollars in summer and $100 to $200 per quarter in winter. If you're paying much more than that, you're using more energy than you have to during non-daylight hours. The trick to even smaller bills for you is energy efficiency, especially in the thermal performance of your home – how you heat and cool it.

I could write a whole book on efficiency. Luckily, I don't have to, because there's a brilliant one already available. If you think your home could be more efficient and you want to shave the last few hundred dollars off your bill, buy this book: *The Energy Freedom Home,* by Beyond Zero Emissions (solarquotes.com.au/efh).

Batteries

Batteries are going to be huge. Storing electricity is the secret to making intermittent renewables (like solar energy) our main source of energy. As explained in Step 4, batteries aren't yet a good economic investment for the typical Australian and they'll increase the carbon footprint of most homes.[16]

This will change in the future when battery prices drop and we can't connect any more renewables to the grid without adding batteries.

When batteries make sense economically or environmentally, I'll let you know. You can subscribe to updates on my blog here: solarquotes.com.au/subscribe.

If you want a battery despite the marginal economics, good on you. You'll be helping the burgeoning home-battery industry and you'll enjoy the peace of mind of blackout protection. Just re-read the advice in Step 5 before you get your quotes.

Cars

Once you've got rid of most of your electricity bill, you should think about your fuel bills.

At the time of writing, I believe we're on the cusp of a revolution with electric cars. If your car has another three or four years left in it, consider hanging on to it a little longer. By then, I'm predicting that electric cars will be a no-brainer for anyone with a big solar system. Beware of buying an expensive new oil burner that no one wants to buy in a few years because everyone wants electric.

---

[16] Solarquotes.com.au/batteryfootprint

CONCLUSION: Beyond Solar

What not to do

In energy, the future is approaching at breakneck speed. You have a choice.

Choice 1: Do nothing

The fact that you've got to the end of this book tells me that you're not a 'do nothing' type of person. You have all the knowledge you need to decide if solar is a good investment for you. The worst case is that you've discovered your roof isn't solar-friendly because there's too much shade or it's a really weird shape.

In that case, I strongly recommend working through the *Energy Freedom Home* book I linked to earlier in this chapter to make your home as efficient as possible. Keep your ear to the ground for 'community solar farms', where you can buy solar panels elsewhere and use their output to credit your bill. They're coming soon.

If you're moving to a new house before the solar will pay for itself, I urge you to make a solar-friendly roof a priority when looking for a new home.

Choice 2: Act now

For the rest of you, with solar-friendly roofs that you'll own for the foreseeable future, get started right now. Go back to Step 2, print out the worksheet and take your first meter reading this Sunday. That's the first small step towards a future where you take responsibility for your energy generation and consumption.

Once you've measured your energy profile, use the free tool I provided to estimate how many panels you can fit on your roof.

That same tool will estimate a financial return for a roof full of solar based on the usage profile you've measured.

Then use one of the methods I suggested to get three quotes for high-quality solar. If you can afford it, err towards the quality end of the hardware choices.

Talk your options over with your installer. Discuss how you'll handle your hot water and where a future battery could go. Make sure the installer's payback projections match the ones you worked out yourself.

If you want to finance the system, avoid the 'no-interest' options and go for a realistic low-rate loan where the repayments are less than the savings.

Then if all the ducks line up, pull the trigger and get solar installed.

The security of low bills for decades, the comfort of cranking the air conditioning and not worrying about the bills, and the satisfaction of offsetting the bad energy choices of our pollies and business leaders all make you feel great. Now you've taken the time to learn about the critical aspects of the energy revolution, you can be confident that you'll benefit from the technology, rather than becoming a victim of energy companies' profiteering or politicians' fearmongering.

A final note

As I write the final words of this book, an episode of *60 Minutes* was just shown on Channel 9 about Australia's energy crisis. In it they interviewed our Federal Minister for Energy, Josh Frydenberg, who revealed that he has large electricity bills every quarter and it

CONCLUSION: Beyond Solar

really puts a dent in his household budget. The presenter asked him if he had solar on his home. He replied that he doesn't because he thinks it's expensive. Apparently, the guy who makes the energy-investment decisions for Australia is not aware that buying good solar on low-cost finance is almost always cheaper than paying electricity bills.

Don't be like Josh. Now you've read this book, you know much more about the practicalities and economics of buying solar than he does. If you ever get interviewed on *60 Minutes* about your electricity bill, instead of looking glum and complaining about a large invoice every three months, you can happily explain that you don't worry about your electricity bills anymore because you did your homework and you have a great value solar system on your roof. A solar system that will provide the security of low bills for decades, no matter how badly the pollies screw up the energy system.

Acknowledgements

Thanks to my Mum, Pirkko Carpenter, for instilling a love of both books and nature, and putting the fire in my belly.

Thanks to Ronald Brakels, for help with the words.

Graphs for the quote section were inspired by Warwick Johnston's excellent PVSell software.

The Author

Finn is a Chartered Electrical Engineer who lives in Adelaide in a solar home made of straw. His last quarterly power bill to power a family of five, a small business and an 8 kW Finnish sauna was a credit of $128.

Finn is the founder of SolarQuotes®, Australia's most popular solar website which, since 2009, has arranged solar quotes for over 350,000 Australians.

Before he founded SolarQuotes®, Finn worked for CSIRO, commercialising their solar and energy efficiency technology.

Find out more at
www.solarquotes.com.au
finn@finnpeacock.com

www.ingramcontent.com/pod-product-compliance
Lightning Source LLC
Chambersburg PA
CBHW070534090426
42735CB00013B/2976